Innovations of Green Process Engineering for Sustainable Energy and Environment

Topical Conference at the 2009 AIChE Annual Meeting

Nashville, Tennessee, USA
8-13 November 2009

ISBN: 978-1-61567-922-5

Printed from e-media with permission by:

Curran Associates, Inc.
57 Morehouse Lane
Red Hook, NY 12571

CURRAN ASSOCIATES INC.
proceedings
.com

Some format issues inherent in the e-media version may also appear in this print version.

Printed by Curran Associates, Inc. (2010)

For permission requests, please contact AIChE
at the address below.

AIChE
3 Park Avenue
New York, NY 10016-5991

Phone: (203) 702-7660
Fax: (203) 775-5177

www.aiche.org

Additional copies of this publication are available from:

Curran Associates, Inc.
57 Morehouse Lane
Red Hook, NY 12571 USA
Phone: 845-758-0400
Fax: 845-758-2634
Email: curran@proceedings.com
Web: www.proceedings.com

Innovations of Green Process Engineering for Sustainable Energy and Environment

Topical Conference at the 2009 AIChE Annual Meeting

Nashville, Tennessee, USA
8-13 November 2009

ISBN: 978-1-61567-922-5

TABLE OF CONTENTS

Industrial Perspectives On Distributed and Compact Chemical Processing

Minquan Cheng, Process Engineering Division, Chevron Energy Technology Company, Houston, TX

World oil demand has exceeded a stunning 85 MMBOPD in recent years. Because of finite nature of fossil resources, International Oil Companies have recently been forced to explore new frontiers such as deepwater and arctic areas to fight world's hunger for energy, and quite impressively, enjoyed success in finding oil and gas. To be able to convert these new resources into reserves, development cost will have to be reduced significantly from its current level. High cost is due to remoteness of asset location, harsh environmental conditions, stringent local content requirements, soaring steel and other commodities price, etc. For high GOR associated oil fields, gas disposal introduces another cost dimension. Venting/flaring of this "byproduct" stream is no longer an option and putting together a pipeline is simply too expensive. In this presentation, deepwater offshore oil and gas processing will be reviewed, and needs for emerging development concepts such as floating LNG, floating GTL, hybrid oil/gas FPSO and their link with distributed/compact processing technology are highlighted. Other emerging energy related processes, such as biomass conversion, waste conversion, and coal to liquids, will also be described and discussed.

See more of this Session: Distributed Chemical Processing Technologies
See more of this Group/Topical: Topical G: Innovations of Green Process Engineering for Sustainable Energy and Environment

Micro-Channel Enabled Process Intensification for Distributed Production of Chemicals and Fuels

Yong Wang, Pacific Northwest National Laboratory, Richland, WA

Growth of energy demand and diversification of feedstock have motivated the development and application of distributed chemical processes for fuel and chemical production. New ways are required to conduct mixing, reaction, separation, energy and material transfer, and process control to break away from the capacity-based scale economy. In this talk, pros and cons of decade-long micro-reaction technology endeavor will be discussed and new technology vision will be presented. These discussions will be elaborated with application examples of hydrogen processing, olefin production, and conversion of natural gas/biomass into liquid fuels.

See more of this Session: Distributed Chemical Processing Technologies
See more of this Group/Topical: Topical G: Innovations of Green Process Engineering for Sustainable Energy and Environment

Carbon-Neutral Transportation Fuels From off-Peak Wind and CO2

David Doty, Doty Scientific, Columbia, SC

Glenn N Doty, Doty Scientific, Columbia, SC

Laura L Holte, Doty Scientific, Columbia, SC

Use of excess off-peak electrical energy to synthesize standard liquid fuels, such as gasoline and jet fuel, could simultaneously address grid stability, domestic oil limitations, climate change, and economic recovery. Simulations have shown that practical innovations should make it possible to reduce CO_2 to CO at over 90% of theoretical efficiency limits (under 1.55 MJ/kg-CO). When combined with our other simulated process advances, it should then be possible to synthesize all hydrocarbons and alcohols from point-source CO_2 and off-peak wind energy by using currently available catalysts at system efficiencies in the range of 53-61%. Off-peak grid energy averaged under $15/MWhr in the MISO hub in the first four months of 2009. (For reference, the cost of energy in gasoline at $3.60/gal is $100/MWhr.) At such prices, synthesized standard liquid fuels (dubbed "WindFuels") could compete even when petroleum is only $45/bbl. There are sufficient amounts of domestic wind resources and point-source CO_2 to produce more than twice our current total transportation fuel usage.

A better alternative for future transportation fuels is needed than those that have previously been advocated - such as biofuels, hydrogen, and methane. When land-use change is properly considered, most biofuels are seen to be only 5% to 20% carbon neutral; and hydrogen has daunting cost challenges with respect to distribution, storage, and end use. It is in our economic and security interests to produce carbon-neutral transportation fuels domestically at the scale of hundreds of billions of gallons annually.

It has long been known that it should theoretically be possible to convert CO_2 and water into standard liquid hydrocarbon and alcohol fuels at high efficiency. The problem has been that prior proposals for doing this conversion have had efficiencies of only 25% to 35% for preferred fuels. That is, the chemical energy in the standard liquid fuels produced (gasoline, ethanol, diesel, etc.) would be about 30% of the input energy required. Our work shows that nearly a factor-of-two gain in efficiency should be possible.

Doty Energy is developing novel processes that will allow the production of fully carbon-neutral standard fuels and plastics from waste CO_2 and off-peak wind or other low-carbon energy at high efficiency and at prices that should soon be competitive with fossil-derived products. Converting CO_2 into fuels can eliminate the need for CO_2 sequestration and reduce global CO_2 emissions by 40%.

The combination of the major technical advances we have simulated over the past two years should permit conversion of CO_2 to fuels to be done at up to 60% efficiency, which is about twice what was expected by most researchers just three years ago. When the input energy is from off-peak wind or nuclear and reasonable consideration is included for climate benefit, WindFuels will sometimes compete even when petroleum is $45/bbl; and most experts are predicting oil will stay over $140/bbl after 2014. Our analysis shows windfuels production per gross land area in good wind regions should exceed biofuels production density in productive farming areas by factors of 4 to 20. Moreover, 99.9% of the land required for the wind farms is generally available for dual use. Switching 70% of global transportation fuels from petroleum to WindFuels should be possible over the next 35 years. The scale-up challenges do not appear to be significantly greater from a technical perspective than similar fuel-synthesis challenges addressed successfully by Germany during WWII.

The WindFuels production system (disclosed in detail in pending patents) will be presented. It is based largely on the commercially proven technologies of wind energy, water electrolysis, and Fischer Tropsch (FT) chemistry. Wind energy is used to split water into hydrogen and oxygen. Some of the hydrogen is used in a process, the so-called reverse water gas shift (RWGS) reaction, that reduces CO2 to carbon monoxide (CO) and water. The CO and the balance of the hydrogen are fed into an FT reactor similar to that commonly used to produce fuels and chemicals from coal or natural gas.

The biggest part of the increase in our projected efficiency comes from improved separations processes in the fully recycled FT loop. Conventional processes for separation of CO2 from the other syngas components have typically required over 6 MJ/kg of CO2. Our simulations indicate a high-pressure cryogenic process can achieve sufficient separation in the FT recycle loop (under 15 molar-%) at under 0.8 MJ/kg. We call the design "full recycle" because essentially all the unreacted H2 and CO are recycled without expansion in the separations processes. To our knowledge, nothing close to full recycle in FTS has been done before. The primary reason it has not been implemented with fossil- or biomass-based FTS is that there is insufficient control flexibility in the H2/CO ratio in the syngas coming from a reforming process to compensate for the variability that will be seen from changes in the water-gas-shift reaction in the FT reactor. Maintaining the desired H2/CO ratio in an RFTS process, on the other hand, is a non-issue, as one has complete and independent control over both the H2 and the CO feed rates in the new syngas.

The next largest gain is expected from an order-of-magnitude advance in cost-effectiveness of gas-to-gas recuperation, which is expected to make 97% effectiveness practical where 75% was previously practical. The third largest gain may be from an optimized RWGS process which promises over 90% efficiency (implying under 1.55 MJ per kg of CO) compared to about 50% for prior demonstrations (about 2.8 MJ/kg-CO). Another significant gain comes from a novel thermodynamic cycle for conversion of the waste heat from the electrolyzer and the FT reactor. A novel approach to RFTS system integration and optimization leads to additional efficiency gains. All the processes have been simulated in detail, and key experiments will soon be carried out to help optimize process conditions.

Windfuels are truly sustainable. The needed CO2 would come from biofuels refineries, ammonia plants, cement factories, ore refining, coal power plants, and other point sources. The water requirements are an order of magnitude less than for biofuels, and the wind will always blow. Windfuels will be over 85% carbon neutral and will flow seamlessly within our current infrastructure.

References

1. Searchinger et al, "Use of U.S. Croplands for Biofuels...", Sci. 319, 1238-1240, 2008.

2. GN Doty, FD Doty, LL Holte, D McCree, S Shevgoor, "...Recycling CO2 into Transportation Fuels - Driving the Off-peak Wind Market", Proceed. WindPower 2009, paper 175, Chicago, 2009.

3. FD Doty, "Hydrocarbon and Alcohol Fuels from Variable, Renewable Energy...," PCT WO 2008/115933, filed 3/2007, http://windfuels.com/PDFs/WO2008115933.pdf.

4. D. Charles, "Corn-Based Ethanol Flunks Key Test", Science 324, p 587, 1 May 2009.

5. SE Park, JS Chang, KW Lee editors, Carbon Dioxide Utilization for Global Sustainability, Studies in Surface Science and Catalysis 153"; OS Joo, KD Jung and J Yonsoo, "CAMERE Process for methanol synthesis from CO2 hydrogenation", pp 67-72, Elsevier, 2004.

6. Mark Z Jacobson, "Review of Solutions to Global Warming, Air Pollutions, and Energy Security", Energy Environ. Sci., 2009.

7. P.L. Spath, D.C. Dayton, (NREL/TP-510-34929, 2003).

8. O.S. Joo, K.D. Jung J. Yonsoo in. Carbon Dioxide Utilization for Global Sustainability, Studies in Surface Science and Catalysis 153, S.E. Park, J.S. Chang, K.W. Lee, Eds., pp 67-72, Elsevier, 2004.

9. M Xiang, D Li, H Qi, W Li, B Zhong, Y Sun, "Mixed alcohols synthesis from CO hydrogenation over K-promoted f"-Mo2C catalysts", Fuel 86, 1298-1303, 2007.

10. R. Zubrin, B. Frankie, T. Kito, AIAA 97-2767, (1997).

11. S Phillips, A Aden, J Jechura, and D Dayton, "Thermochemical Ethanol via Indirect Gasification and Mixed Alcohol Synthesis...", NREL/TP-510-41168, 2007.

12. FD Doty and S Shevgoor, "Compact, High-Effectiveness, Gas-to-gas Compound Recuperator with Liquid Intermediary (CRLI)", HT2009-88372, ASME Joint Conferences, San Francisco, 2009.

13. FD Doty and S Shevgoor, "A Dual-source Organic Rankine Cycle (DORC) ..." ES2009-90220, ASME Joint Conferences, San Francisco, 2009.

See more of this Session: Distributed Chemical Processing Technologies
See more of this Group/Topical: Topical G: Innovations of Green Process Engineering for Sustainable Energy and Environment

Synergistic Routes to Liquid Fuel for a Petroleum Deprived Future

Navneet R. Singh, School of Chemical Engineering and Energy Center at Discovery Park, Purdue University, West Lafayette, IN

Rakesh Agrawal, School of Chemical Engineering and Energy Center at Discovery Park, Purdue University, West Lafayette, IN

As compared to biomass gasification/Fischer-Tropsch synthesis, hydropyrolysis/hydrodeoxygenation (HDO) based processes have a potential to achieve high biomass carbon conversion to liquid fuel with much lower amounts of supplementary H2. On the basis of this observation, we suggest a Hydrogen Bio-oil (H2Bioil) process using fast-hydropyrolysis/HDO that has a potential to produce nearly double the amount of liquid fuel when compared to the existing biofuel processes while requiring only modest quantities of supplementary H2. The optimal operating mode for the H2Bioil process is suggested to be in an entrained bed mode in presence of H2 with gas phase HDO of hydropyrolyzed vapors. A remarkable result due to reduced need for the supplementary H2 is that it provides synergistic integration of the H2Bioil process with a coal gasification power plant or a small scale steam natural gas reformer leading to a dramatic increase in the liquid fuel production from biomass and coal or natural gas. Here, hot syngas (T>500°C) from a coal gasifier or methane reformer supplies H2/CO for hydropyrolysis and deoxygenation as well as heat for the process. This result is exciting because it presents us with an option to build integrated H2Bioil processes sooner rather than later when the cost-effective H2 becomes available from a carbon-free energy source such as solar or nuclear. The H2Bioil process and its integrated version with a small scale natural gas reformer have strong potential to be attractive on a small scale while being more efficient than any current biomass to liquid fuel process in operation.

See more of this Session: Distributed Chemical Processing Technologies
See more of this Group/Topical: Topical G: Innovations of Green Process Engineering for Sustainable Energy and Environment

NOVEL MICROFIBROUS COMPOSITE BED REACTOR INCORPORATED WITH SILVER CATALYST FOR HIGH EFFICIENCY GAS-PHASE OXIDATION OF ALCOHOLS

*Jiping Mao, Miaomiao Deng, Yong Lu**
Shanghai Key Laboratory of Green Chemistry and Chemical Processes, Department of Chemistry, East China Normal University, Shanghai, China

Introduction

Selective oxidation of alcohols to the corresponding carbonyl compounds is one of the most important chemical transformations in organic chemistry for both laboratory and industrial manufacturing. Carbonyl compounds such as aldehydes and ketones are the valuable synthetic intermediates for many drugs, vitamins and fragrances.[1-3] Traditionally, such syntheses are mostly preformed in the presence of large amount of volatile organic solvents (*e.g.*, ethanol, benzene, *p*-xylene, etc.) with stoichiometric inorganic oxidants, notably Cr reagents.[4-7] In nearly two decades, various catalytic systems using environmentally benign oxidants like molecular oxygen, hydrogen peroxide, organic peroxides under homogeneous (or heterogeneous) conditions have been developed,[3,5-13] aiming at replacing the traditional toxic and expensive stoichiometric oxidation processes. Among them, the gas-phase selective oxidation of alcohols shows great potential to develop efficient, cost-effective and green processes for the production of carbonyl compounds.[8-13] It is desirable to significantly intensify the reaction heat removal from the catalyst bed, and meanwhile, to suppress the thermal cracking and over oxidation of the alcohols (especially the large molecules). A significant challenge facing this effort is the development of high-performance catalyst and/or new-type reactor with unique combination of excellent heat conductivity and high low-temperature activity/selectivity.

Some silver-based catalysts including electrolytic silver and supported silver catalysts, have been demonstrated recently for the selective oxidation of alcohols[8-13] in the packed bed reactor. However, some serious problems still exist in the applications of such catalysts. For example, the electrolytic silver catalyst, widely used for methanol oxidation to produce aldehyde using thin-bed reactor, has good thermal conductivity but is not active at below 500 ^{o}C or not selective at above 500 ^{o}C for the gas-phase oxidation of larger alcohols such as benzyl alcohol.[8,10] The supported silver catalysts using the carriers such as Al_2O_3, SiO_2 and SiC, have better catalytic activity at relatively low temperatures in the packed bed reactor.[9] However, their weak thermal conductivity easily generates partial hotspots in reaction beds for the strongly exothermic gas-phase selective oxidation of alcohols. This is not only a main cause of the catalyst degradation but also a hidden danger. More recently, the Au/SiO_2 catalysts with or without copper additives have been reported for the gas-phase selective oxidation of benzyl alcohol.[14] These Au-beased catalysts delivered excellent low-temperature activity with very high target product selectivity. Nevertheless, bed-temperature increase of >40 ^{o}C was observed even in a lab-scale micro-reactor owing to the poor thermal conductivity of the catalyst bed.[14] For the reasons given above, their practical applications are very limited.

Hence, it is important to render a new kind of catalyst and/or new-type reactor with not only good low-temperature activity/selectivity but also significantly intensified intrabed heat transfer. To accomplish this goal, a novel silver catalyst supported on LTA zeolite film [8] coated on a copper grid has been reported and demonstrates good low-temperature activity with high selectivity to benzaldehyde. Recently, a new class of microfibrous carriers using microfibers has been invented by Tatarchuk.[15,16] This approach has been extensively applied to develop high-efficiency miniature H_2 generator for fuel cell power supplies.[17-20] This microfibrous structure provides large void volume, entirely open structure, large surface-to-volume ratio, high permeability, high thermal conductivity and unique form factors. For applications that entrap small particles within or load active components (e.g., metals) on the microfibrous network, unique combinations of surface area, pore size/particle size, thermal conductivity, and large void volume are obtained. This approach permits high efficiency process and advanced design of reactor with many beneficial properties that avoid the frustrated problems encountered in the conventional approaches (*e.g.,* conventional packed beds). Our previous microfibrous composites with entrapment of preferential oxidation (PROX) CO catalyst particulates and with entrapment of H_2S sorbent particulates for hydrogen fuel cleanup in proton exchange membrane fuel cell (PEMFC) applications do both provide a 3-fold or more promotion of bed utilization efficiency.[17-20] These approaches also lead to significant reduction of the overall reaction bed weight and volume compared to the packed beds with 1-2 mm dia. catalyst/sorbent pellets.[17-20] More interestingly, such microfibrous structure, with unique form factors, can be made into thin sheets (from submillimeter to several millimetres in thickness) of large area and/or pleated sheet structure to control pressure drop and contacting efficiency in a beneficial manner different from other traditionally employed contacting schemes including packed beds, fluid beds, honeycomb monoliths or wovens.[17,18]

Accordingly, it is safe to say that the microfibrous structure is an excellent and promising technology for the development of high-performance catalysts (or new-type microfibous bed reactor), especially for the strongly endothermic or exothermic reactions. Metallic network structure definitely possesses the excellent thermal conductivity. For the strongly exothermic gas-phase oxidation of alcohols, the excellent heat conductivity of the catalytic bed facilitates the isothermal operation thereby likely ensuring the activity/selectivity maintenance. Nonporous structure of the microfibers results in the high accessibility of active components deposited on. Actually, the large void volume and the open structure as well as the excellent heat conductivity are the central notion in promoting the steady-state volumetric reaction rate.[21] Exactly, silver catalyst supported on a sinter-locked Ni-fiber has been reported and this novel microfibrous bed reactor indeed demonstrates good low-temperature activity for the gas-phase selective oxidation of benzyl alcohol with high target product selectivity.[10] However, the nature for their low-temperature activity as well as the reactivity of this new-type micro-structured reactor for the other alcohols is still not clear.

In present study, sinter-locked three-dimensional microfibrous network consisting of 5 vol% 8 μm (dia.) Ni fibers or 12 μm (dia.) SS-316L fibers and 95 vol% void volume were utilized to support Ag by incipient wetness impregnation with $AgNO_3$. The reactivity of the micro-structured catalytic bed incorporated with silver catalysts was measured in the case of gas-phase selective oxidation of mono-/di-alcohols and aromatic alcohols using O_2 as oxidants. The microfibrous Ag/Ni-fiber demonstrated much higher low-temperature activity/selectivity at 330-380 °C and atmospheric pressure, compared to the packed bed with the electrolytic silver catalyst. To gain insight into the nature for the significant improvement of low-temperature activity, the catalytically relevant physiochemical properties of the microfibrous-structured silver catalysts were studied by X-ray diffraction (XRD), O_2-chemisorption, H_2-temperature programmed reduction (H_2-TPR), O_2-temperature programmed oxidation (O_2-TPO), X-ray photoelectron spectroscopy (XPS), scanning electron microscope (SEM), and UV-visible diffuse reflectance spectroscopy (UV-vis DRS).

Experimental Section

Preparation of microfibrous-structured silver catalyst

The lab-scale thin-sheet microfibrous metal networks were built up by the wet-lay papermaking process[20] using 2-3 mm microfiber chops of Ni (8 μm diameter) and stainless steel 316L (SS-316L, 12 μm diameter), followed by the high-temperature sintering in hydrogen atmosphere. The detailed procedure for producing the microfibrous metal network is described as following: 5 g of metal fibers and 1.7 g of cellulose fibers were added into water (1 L) and stirred vigorously to produce a uniform suspension. The resulting suspension was transferred into the head box of a 159 mm dia. circular sheet former (ZCX-159A, made in China) filled with 8 L water under manual mixing. A 159 mm circular preform was then formed by draining and drying in air. As-made preform paper was oxidized in air at 450 °C to remove the cellulosic binders and subsequently sintered in hydrogen at 950 °C for Ni fibers and at 1100 °C for SS-316L fibers to create the three-dimensional sinter-locked networks. The resulting microfibrous networks were incipiently impregnated with the aqueous solution containing appointed amount of $AgNO_3$, dried overnight at 110 °C and calcined in air at temperature range between 400 and 700 °C for 4 h to obtain the microfibrous-structured silver catalysts. For comparison, electrolytic silver (supplied by the HePing noble metal catalyst company) was crushed and 200-300 μm particulates were collected for reaction use. Table 1 summarizes the characteristics of the typical catalysts studied.

Table 1. Characteristics of the Typical Catalysts Studied.

Catalyst	**Ag/Ni-fiber**	**Ag/Steel-fiber**	**Electrolytic silver***
Support	Ni-fiber	SS-fiber	-
Ag loading (wt%)	10.1	9.8	99.9
Void volume fraction (%)	95	95	43.5
Density ($g.cm^{-3}$)	0.33	0.33	3.5

* Particle size: 200-300 μm.

Characterization

XRD patterns were obtained with a Bruker D8 ADVANCE diffractometer over a 2θ range of 20-85° in step of 0.02°, using Cu Kα radiation at an acceleration voltage of 40 kV. UV-vis DRS spectra were recorded in the range of 200-500 nm at room temperature using a Shimadzu UV-2400PC spectrophotometer. SEM images and Ag elemental mapping images of the catalysts were obtained on a Hitachi S4800 instrument equipped with an energy dispersive

X-ray analyses (EDX) unit (Oxford, UK). XPS data were obtained on a Versaprobe Phi 5000 ULVAC–PHI instrument using Al Kα radiation ($h\nu$ = 1486.6 eV). The adventitious C1s line (binding energy of 284.6 eV) was used as reference for correcting charge shift.

The experiments on H_2-TPR, O_2-TPO and O_2-chemisorption were carried out on a Quantachrome ChemBET 3000 chemisorption apparatus with a TCD. The sample loading was 0.1 g for each trial. The mixture of 5% H_2/Ar and mixture of 3% O_2/He were employed as carrier gases for both H_2-TPR and O_2-TPO tests, respectively. In these cases, the carrier gas flow rate was maintained at 40 mL/min while the temperature of sample was ramped at 10 ^{o}C/min. Note that the samples after the H_2-TPR tests were cooled down to room temperature in a He flow and subsequently used for the O_2-TPO tests. O_2-chemisorption experiments were carried out at 200 ^{o}C to determine the surface Ag atoms by assuming an O/Ag ratio of 1, according to Gavriilidis et al.[22] The samples were pre-reduced with 5% H_2/Ar at 300 ^{o}C for 2 h and cooled down to 200 ^{o}C in He flow. Subsequently, a sample of 3% O_2/He was pulsed, with a pulse volume of 100 μL, to the reactor every 5 min until the O_2 peak intensity remained unchanged.

Reactivity tests

The gas-phase selective oxidation of alcohols on these catalysts with molecular oxygen was carried out on a fixed-bed reactor under atmospheric pressure. The reactor was a quartz tube with an inner diameter (i.d.) of 16 mm. The catalysts were all loaded and directly exposed to the reactant stream without any pretreatment. The catalyst loading was 0.3 g for the microfibrous-structured silver catalysts but was 1.0 g for the electrolytic silver. For the microfibrous-structured silver catalysts, the circular chips (16.1 mm diameter) were punched down from their large sheet sample and packed layer-up-layer into the tube reactor. Note that the diameter of 0.1 mm larger than the i.d. of the tubular reactor was retained deliberately to avoid the appearance of the gap between the reactor wall and the edges of the catalyst chips thereby preventing the gas bypassing. Alcohols were fed continuously using a high-performance liquid pump, in parallel with O_2 (oxidant) and N_2 (diluted gas) feeding using the calibrated mass flow controllers, into the reactor heated to the desired reaction temperature. The reaction temperature, weight hourly space velocity (WHSV) and molar ratio of O_2 to alcoholic hydroxyl (O_2/ol) were varied in range from 300 to 500 ^{o}C, 4 to 50 h^{-1}, and 0.4 to 1.0, respectively. The effluent was cooled using an ice-salt bath (-15 ^{o}C) to liquefy the condensable vapors for analyzing by an HP 5890 gas chromatography-flame ionization detector (GC-FID) with a 60-m HP-5ms capillary column. The gas-phase products such as CO_2, CO and C1-C3 hydrocarbons were analyzed by an HP-5890 GC with thermal conductivity detector (TCD) and a 30-m AT-plot 300 capillary column.

Results and Discussion

Microfibrous structure and surface morphology

Figure 1 shows the microstructure, surface morphology and silver distribution of the microfibrous media. Figure 1a and Figure 1d show the SEM images of the microfibrous structure and surface morphology of the sintered microfiber products before loading silver. Clearly, the sintering process led to a sinter-locked three-dimensional network while the large void volume and entirely open structure remained. Figure 1b and Figure 1e show the surface morphology of the microfibrous structured Ag catalysts. The fiber surface became much rougher after loading Ag by impregnation method with the aqueous solution of silver nitrate. In comparison with the neat microfibers (Figure 1a and Figure 1d), a number of particles appeared on the catalysts and the particle size was smaller for 10 wt% Ag/Ni-fiber (Figure 1b) than that for 10 wt% Ag/SS-fiber (Figure 1e). Figure 1c and Figure 1f show the EDX element mapping images of Ag of the corresponding samples in Figure 1b and Figure 1e, respectively. It is clearly that the Ag distribution by EDX was well consistent with the observation of the particles presented on the catalysts. This indicates that the particles as shown in Figure 1b and Figure 1e were almost made up of Ag element and Ag was dispersed better on the Ni microfibers than on the SS-316L microfibers. Characteristic XRD peaks ascribed to Ag(111) (2θ: 38.1°), Ag(200) (2θ: 44.3°), Ag(220) (2θ: 64.4°), and Ag(311) (2θ: 77.5°) were clearly observed on the both microfibrous-structure supported silver catalyst samples, indicating the silver in such two catalysts was mainly existed in the metallic state.

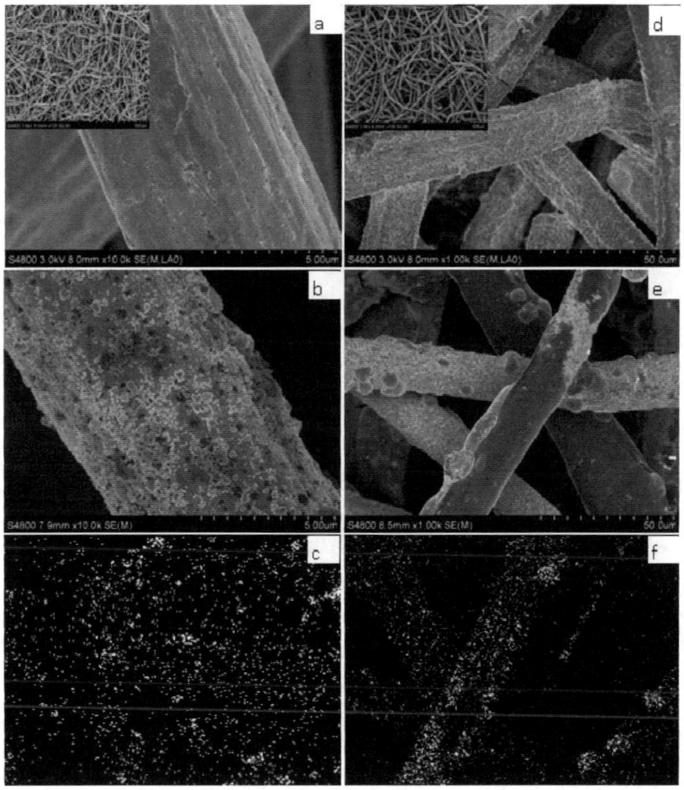

Figure 1. Morphology and microfibrous structure. SEM images of sinter-locked 8-μm-Ni-fiber before (a) and after (b) loading 10 wt% Ag (10 wt% Ag/Ni-fiber), SEM images of sinter-locked 12-μm-SS-fiber before (d) and after (e) loading 10 wt% Ag (10 wt% Ag/SS-fiber). EDX Ag element spectra of 10 wt% Ag/Ni-fiber (c) and 10 wt% Ag/SS-fiber (f).

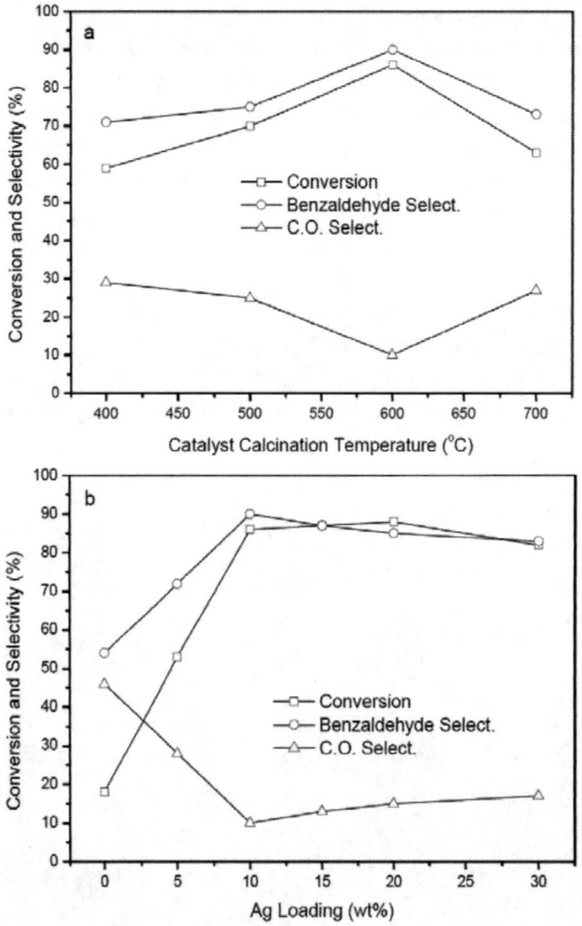

Figure 2. Effects of catalyst calcination temperature (a) and Ag loading (b) on the performance of Ag/Ni-fiber catalysts for gas-phase oxidation of benzyl alcohol at 380 °C. WHSV = 25 h^{-1}, molar ratio of O_2 to alcoholic hydroxyl (O_2/ol) = 0.6.

Catalytic performance

According to the different molecular structures or numbers of hydroxyl groups, alcohols can be divided into aromatic-alcohols, mono-alcohols and di-alcohols. Therefore, we selected benzyl alcohol, 1-butanol and 1,2-propylene glycol as the representatives of the different kind of alcohols for systematic tests in this study. Of course, extended tests for other alcohols including octanol, cyclohexanol and phenethyl alcohol were also performed to examine the generality of our microfibrous-structured silver catalysts.

Gas-phase selective oxidation of aromatic and alicyclic alcohols: **Effects of Ag loading and catalyst calcination temperature**. Figure 2 shows the effects of Ag loading and catalyst calcination temperature on the performance of the Ag/Ni-fiber catalysts for gas-phase selective oxidation of benzyl alcohol at 380 °C using a WHSV of 25 h^{-1}. As we can see, the reactivity of the Ag/Ni-fiber catalysts was greatly dependent on the Ag loading and catalyst calcination temperature under the compatible reaction conditions. For the Ag/Ni-fiber catalysts with Ag loading of 10 wt%, visible increase of benzyl alcohol conversion and selectivity to benzaldehyde was observed as the calcination temperature was increased from 400 to 600 °C: 59% vs. 86% for conversion and 71% vs. 90% for selectivity to benzaldehyde (Figure 2a). As noted in previous work, Ag^{+} ions and $Ag_n^{\delta+}$ clusters are active and selective for the gas-phase oxidation of alcohols.[8,23-26] Calcination at 600 °C likely facilitated the formation of such active/selective sites on the surface of Ag/Ni-fiber catalysts thereby leading to good performance. Furthermore, some kind of new interaction between Ag and Ni-fiber at their interface was likely formed after calcination at 600 °C. This was believed to further promote the activity/selectivity of the Ag/Ni-fiber catalysts since poor conversion (63%) and benzaldehyde selectivity (81%) were obtained when the sinter-locked SS-316L microfiber was used to replace the Ni-fiber carrier only. Further increase of the calcination temperature up to 700 °C, however, significantly degraded both the conversion and selectivity of the Ag/Ni-fiber catalysts (Figure 2a), likely due to the severe sintering of Ag as well as the deep oxidation of the Ni-fiber. The selectivity to byproducts (*e.g.*, CO_2, benzene and toluene) showed a reverse tendency of the conversion and the benzaldehyde selectivity with the increase of the catalyst calcination temperature. For reference, the neat Ni-fiber carrier (i.e., Ag loading of 0 wt%) with pretreatment at 600 °C in air was also tested and induced a conversion of 18% with a benzaldehyde selectivity of 54% only (Figure 2a).

To address the effect of Ag loading, the calcination temperature remained at 600 °C for catalyst preparation. As shown in Figure 2b, Ag/Ni-fiber with the Ag loading of 10 wt% was the best catalyst for the selective oxidation of benzyl alcohol. In comparison with the neat Ni-fiber, the Ag loading of 5 wt% resulted in at once a significant increase of the conversion and selectivity. Increasing the Ag loading from 5 to 10 wt% greatly promoted the conversion and selectivity again. Further increasing the Ag loading to 20 wt% almost made no impact on the conversion but slightly degraded the selectivity to benzaldehyde. Further increasing the Ag loading up to 30 wt% visibly deteriorated both the conversion and selectivity.

Therefore, in the ongoing test experiments, the microfibrous structured silver catalysts with Ag loading of 10 wt% and catalyst calcination temperature of 600 °C were employed if not specified.

Gas-phase selective oxidation of aromatic and alicyclic alcohols: **Effects of reaction conditions**. The effects of reaction conditions on the performance of Ag/Ni-fiber catalyst for the gas-phase selective oxidation of benzyl alcohol were carefully investigated, with the results as compiled in Table 2. It is clear that the optimal reaction temperature was 380 °C with the WHSV of 30 h^{-1} and O_2/ol of 0.6; good conversion of 85% with the highest benzaldehyde selectivity of 94% was obtained, corresponding to the best target product yield of 80%. When the reaction temperature was increased from 330 to 420 °C using WHSV of 20 h^{-1} and O_2/ol of 0.6, the conversion of benzyl alcohol was increased from 57% to 99% but the best selectivity of 86% was presented at 380 °C. At 380 °C and WHSV of 20 h^{-1}, increasing O_2/ol from 0.4 to 0.6 obviously promoted the conversion but degraded the target product selectivity a little; further increasing the O_2/ol from 0.6 to 1.0 provided a little increase in conversion but obviously deteriorated the selectivity to benzaldehyde. At 380 °C and O_2/ol of 0.6, conversion of benzyl alcohol was smoothly decreased from 97% to 63% with the increase of WHSV from 10 to 50

h^{-1}, but the selectivity to benzaldehyde reached maximum (~94%) at the WHSV of 30 h^{-1} and almost was retained at this point with further increasing the WHSV up to 50 h^{-1}.

Table 2. Effects of reaction conditions on the performance of microfibrous-structured silver and electrolytic silver catalysts for gas-phase oxidation of benzyl alcohol [a]

Catalyst [b]	**WHSV** (h^{-1})	**O_2/ol**	**Reaction Temp.** (°C)	**Conversion** (%)	**Selectivity** (%)	
					BzH [d]	**C.O.** [e]
Ag/Ni-fiber [c]	10	0.6	380	97	81	19
	20	0.6	380	92	86	14
	30	0.6	380	85	94	6
	40	0.6	380	76	92	8
	50	0.6	380	63	87	13
	20	0.4	380	79	90	10
	20	0.8	380	93	79	21
	20	1.0	380	95	73	27
	20	0.6	330	57	81	19
	20	0.6	350	73	82	18
	20	0.6	420	99	70	30
Ag/SS-fiber [c]	10	0.6	450	92	84	16
	20	0.6	450	88	91	9
	30	0.6	450	85	90	10
	20	0.4	450	62	89	11
	20	0.8	450	91	79	21
	20	0.6	390	61	92	8
	20	0.6	480	90	71	29
Electrolytic	8	0.6	500	68	62	38
silver	20	0.6	500	55	79	21

[a] Each reaction condition was run for 3 h with mass balance of 93-96%, during which the experimental data were collected. [b] The catalyst loading was 0.3 g for the microfibrous-structured silver catalysts but was 1.0 g for the electrolytic silver. [c] Ag loading: 10 wt% (error: ±0.3 wt% by ICP (IRIS Intrepidll XSP)), calcination condition: 600 °C in air for 4 h. [d] BzH = Benzaldehyde. [e] C.O. = cracking and over oxidation byproducts.

Gas-phase selective oxidation of aromatic and alicyclic alcohols: **Comparison**. Test experiments on the gas-phase selective oxidation of benzyl alcohol on the Ag/SS-fiber and electrolytic silver catalysts were performed for comparison. The reaction results are also shown in Table 2. Clearly, the 10 wt% Ag/Ni-fiber catalysts delivered much higher conversion and selectivity compared to the electrolytic silver catalysts, even operating the reaction at much lower reaction temperature (380 °C vs. 500 °C) with quite higher WHSV (30 h^{-1} vs. 8 h^{-1}). The 10 wt% Ag/SS-fiber did not yet yield better low-temperature activity/selectivity than the 10 wt% Ag/Ni-fiber throughout the entire reaction parameter range studied here. Note that the target product yield of 80% was also achievable at 450 °C.

It was estimated that the Ag/Ni-fiber catalysts provided at least 4-fold promotion of the catalytic bed utilization efficiency (*i.e.*, much higher WHSV: 40 h^{-1} for Ag/Ni-fiber vs. 8 h^{-1})

compared to the electrolytic silver, on the base of the equivalent alcohol conversions: 76% for Ag/Ni-fiber at 380 °C; 68% for electrolytic silver at 500 °C. Moreover, this new approach permitted dramatic saving of Ag: the microfibrous bed of 10 wt% Ag/Ni-fiber catalyst (0.3 g) has only 0.03 g Ag, only 3% that of the electrolytic silver (1.0 g in present work).

Test experiments on the gas-phase oxidation of phenethyl alcohol and cyclohexanol for both the 10 wt% Ag/Ni-fiber and electrolytic silver catalysts were also performed with the results as shown in Figure 3. Our 10 wt% Ag/Ni-fiber catalyst exhibited much higher conversion and selectivity to the corresponding target products even at quite low temperature with very high WHSV, compared to the electrolytic silver catalyst. Clearly, this microfibrous structured catalyst demonstrated highly effective and efficient gas-phase oxidation of aromatic and alicyclic alcohols, with significantly promoted low-temperature reactivity and multi-fold amplification of the reactor throughput.

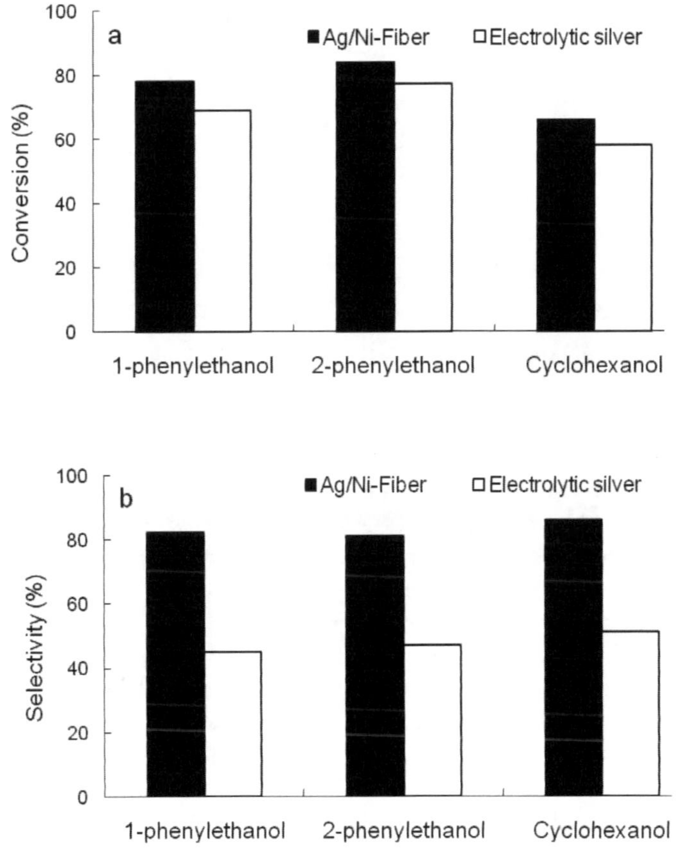

Figure 3. Comparison of conversion (a) and selectivity to carbonyl products (b) for gas-phase oxidation of various aromatic and alicyclic alcohols over 10 wt% Ag/Ni-fiber at 380 °C and electrolytic silver at 500 °C. O_2/ol = 0.6, WHSV = 20 h^{-1}.

Table 3. Effects of reaction conditions on the performance of microfibrous-structured silver and electrolytic silver catalysts for gas-phase oxidation of n-butanol [a]

Catalyst [b]	WHSV (h^{-1})	O_2/ol	Reaction Temp. (°C)	Conversion (%)	Selectivity (%)	
					B.A. [d]	C.O. [e]
Ag/Ni-fiber [c]	10	0.6	380	77	80	20
	15	0.6	380	73	85	15
	20	0.6	380	68	89	11
	30	0.6	380	52	84	16
	40	0.6	380	46	79	21
	20	0.4	380	59	94	6
	20	0.8	380	61	82	18
	20	0.6	330	52	93	7
Ag/SS-fiber [c]	10	0.6	450	78	79	21
	20	0.6	450	71	87	13
	30	0.6	450	63	83	17
	40	0.6	450	57	78	22
	20	0.4	450	51	64	36
	20	0.8	450	80	80	20
	20	0.6	390	61	79	21
	20	0.6	480	74	82	18
Electrolytic silver	8	0.6	500	61	32	68
	20	0.6	500	43	60	40

[a] See footnote ***a*** in Table 2. [b] See footnote ***b*** in Table 2. [c] See footnote ***c*** in Table 2. [d] B.A. = n-butyl aldehyde. [e] See footnote ***e*** in Table 2.

Gas-phase selective oxidation of acyclic mono-alcohols. Table 3 shows the conversion and product selectivity for the gas-phase selective oxidation of n-butanol, using the 10 wt% Ag/Ni-fiber, 10 wt% Ag/SS-fiber and electrolytic silver catalysts. As in the case of benzyl alcohol, the conversion and product selectivity were dependent on the catalysts and significantly affected by the WHSV, O_2/ol and reaction temperature. The Ag/Ni-fiber and Ag/SS-fiber catalysts were more active and selective in the gas-phase oxidation of n-butanol to n-butylaldehyde under relatively mild reaction conditions, compared to the electrolytic silver catalyst. Likewise, the 10 wt% Ag/SS-fiber catalyst had not yield the activity/selectivity better than that for the 10 wt% Ag/Ni-fiber yet, being further confirmed the existence of special Ag-Ni interaction that is contributed to the good low-temperature activity/selectivity. In addition, with the increase of WHSV, O_2/ol and reaction temperature, the conversion and selectivity evolution for the selective oxidation of butanol was alike with that for the selective oxidation of benzyl alcohol. We also proceeded to investigate the effects of Ag loading and catalyst calcination temperature on the performance of Ag/Ni-fiber catalyst for the selective oxidation of n-butanol, indicating that the 10 wt% Ag/Ni-fiber calcined at 600 °C still was the best catalyst.

Figure 4 shows the conversion and target product selectivity for the selective oxidation of 1-octanol and 2-octanol using 10wt% Ag/Ni-fiber and electrolytic silver catalysts. Clearly, the traditional electrolytic silver catalyst had not yielded better activity for all alcohols than our microfibrous catalyst yet, even operating at 500 °C. In addition, operation at elevated temperature facilitated the thermal cracking and over oxidation processes thereby leading to poor selectivity for the electrolytic silver catalyst. In contrast, the good low-temperature activity of the Ag/Ni-fiber catalyst ensured the high conversion while effectively suppressing the thermal cracking and over oxidized products generated at elevated temperatures, and therefore, a higher yield of the target aldehydes could be easily obtained. Moreover, multi-fold increase of the WHSV resulted in a high yielding rate of the target aldehydes.

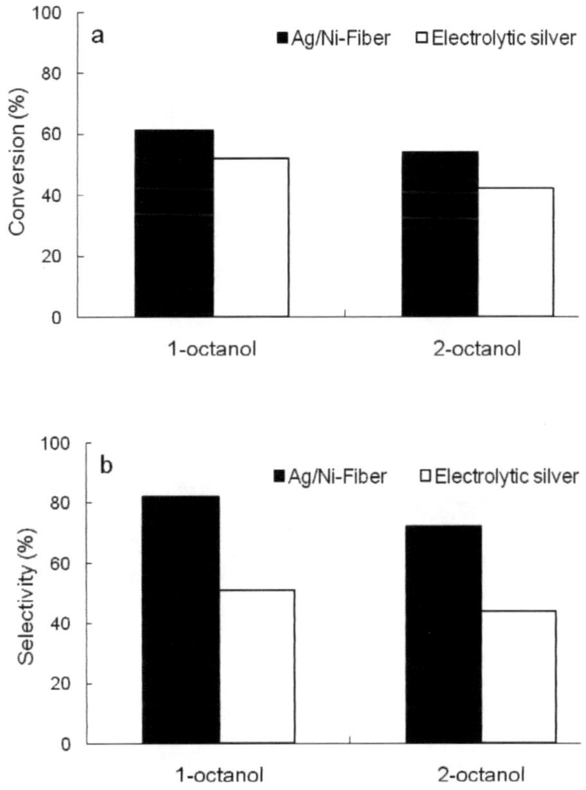

Figure 4. Comparison of conversion (a) and selectivity to carbonyl products (b) for gas-phase oxidation of various acyclic alcohols over 10 wt% Ag/Ni-fiber at 380 °C and electrolytic silver at 500 °C. O_2/ol = 0.6, WHSV = 20 h^{-1}.

Gas-phase selective oxidation of 1,2-propylene glycol. In comparison with the above mono-/aromatic-alcohols, the selective oxidation of di-alcohols is particularly challenging due to the one more hydroxyl group. Take for instance the selective oxidation of 1,2-propylene glycol of which the target product is methyl glyoxal. However, lean-oxidation to form hydroxyketone is likely superior to the moderate oxidation. Because of the existence of the secondary hydroxyl group, the 10 wt% Ag/Ni-fiber catalyst calcined at 600 °C is the best one for the mono-/aromatic-alcohols but likely not for the 1,2-propylene glycol. Therefore, the effects of Ag loading and catalyst calcination temperature on the performance of Ag/Ni-fiber catalyst for this reaction were investigated firstly. It was found that the optimal catalyst calcination temperature was still 600 °C but the best Ag loading was changed to 20 wt%.

Table 4 shows the conversion and product selectivity for the selective oxidation of 1,2-propylene glycol catalyzed by the 20 wt% Ag/Ni-fiber and the electrolytic silver catalysts. Our microfibrous catalyst provided excellent low-temperature activity with quite high selectivity to methyl glyoxal compared to the electrolytic silver catalyst. At a conversion of 97-99%, the 20 wt% Ag/Ni-fiber delivered the best target product selectivity (83%, two times as high as that (37%) of the electrolytic silver) while leading to a decrease in the reaction temperature by 120 °C and a one-fold increase of WHSV. In addition, the 10 wt% Ag/Ni-fiber catalyst delivered a conversion of only ~20% at 330 °C using a O_2/ol of 0.8 and WHSV of 8 h^{-1}, about one fifth that of the 20 wt% Ag/Ni-fiber catalyst. The reason for this is not clear yet.

Table 4. Effects of reaction conditions on the performance of microfibrous-structured silver and electrolytic silver catalysts for gas-phase oxidation of 1,2-propylene glycol [a]

Catalyst [b]	**WHSV** (h^{-1})	**O_2/ol**	**Reaction Temp.** (°C)	**Conversion** (%)	**Selectivity** (%)		
					H.P. [d]	**M.G.** [e]	**C.O.** [f]
Ag/Ni-fiber [c]	15	0.6	330	93	22	70	8
	15	0.7	330	96	17	72	11
	15	0.8	330	99	8	83	9
	15	0.9	330	99	7	67	26
	15	0.6	300	63	28	58	14
	15	0.6	360	96	23	48	29
	8	0.6	330	97	18	58	24
	25	0.6	330	87	27	64	9
Ag/SS-fiber [c]	15	0.6	380	97	33	50	17
	15	0.7	380	98	24	55	21
	15	0.8	380	98	20	47	33
	15	0.7	360	91	27	46	27
	15	0.7	400	99	18	36	46
	8	0.7	380	99	19	46	35
	25	0.7	380	89	28	53	19
Electrolytic silver	4	0.8	450	97	23	37	40
	8	0.8	450	90	29	42	29

[a] See footnote ***a*** in Table 2. [b] See footnote ***b*** in Table 2. [c] Ag loading: 20 wt% (error: ±0.3 wt% by ICP (IRIS Intrepidll XSP)), calcination condition: 600 °C in air for 4 h. [d] H.P. = 1-hydroxy-2-propanone. [e] M.G. = methyl glyoxal. [f] See footnote ***e*** in Table 2.

O_2 chemisorption: silver dispersion

It is widely accepted that highly dispersed metal particles show high catalytic activity due to a high fraction of surface metal atoms. We asked ourselves if the Ag dispersion of these catalysts was different thereby leading to the difference in their reactivity evolution. To make clarity, Ag dispersion of the catalyst samples was determined using O_2 chemisorption at 200 °C with results as summarized in Table 5. According to Gavriilidis et al.,[22] the supported silver catalysts can reach the mono-layer saturated chemisorption of O atom at 200 °C. As shown in Table 5, the supported Ag catalyst samples delivered equivalent surface Ag atoms; the electrolytic silver had surface Ag atoms two times as high as the supported silver catalysts.

For comparison, the turnover frequencies (TOFs) to benzyl alcohol were calculated by normalizing the benzyl alcohol conversion by the number of surface Ag atoms per gram catalyst. At equivalent benzyl alcohol conversion (60-70%) at O_2/ol of 0.6, the TOFs were: 57 s^{-1} for 10 wt% Ag/Ni-fiber under 380 °C and WHSV of 50 h^{-1}, 26 s^{-1} for 10wt% Ag/SS-fiber under 450 °C and WHSV of 20 h^{-1}, and 13 s^{-1} for electrolytic silver under 500 °C and WHSV of 10 h^{-1}. Clearly, no correlation between activity of the catalysts and their surface Ag atoms could be established, likely suggesting that the difference in the electronic properties of the catalyst samples might be the main cause for their quite different catalytic performance.

Table 5. Results of O_2-chemisorption at 200 °C.

Catalyst	Ag/Ni-fiber [a]	Ag/SS-fiber [a]	Electrolytic silver
O_2 in-take (μmol/$g_{cat.}$)	1.25	1.08	2.22
Surface Ag atoms per gram catalyst (×10^{18})	1.38	1.20	2.45

[a] Ag loading: 10 wt%, calcination condition: 600 °C in air for 4 h.

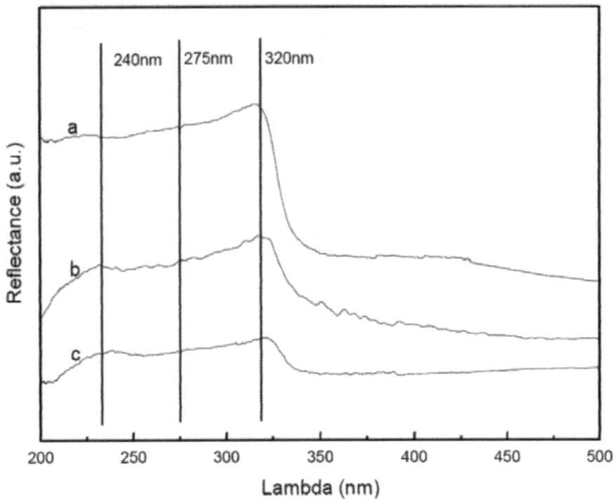

Figure 5. UV–vis DRS spectra of electrolytic silver (a, after calcination at 600 °C in air for 2 h), 10 wt% Ag/Ni-fiber (b), and 10 wt% Ag/SS-fiber (c).

UV-vis DRS and XPS: Active sites

As generally accepted, Ag^+ ions and $Ag_n^{\delta+}$ clusters are active and selective for the gas-phase oxidation of alcohol.[8,23-26] Is it possible that the catalyst samples had quite difference in the active site (Ag^+ ions and $Ag_n^{\delta+}$ clusters) concentration thereby leading to the different catalytic activity? To answer this question, the chemical state of silver particles in the silver-based catalysts was analyzed using UV–vis DRS technique. Figure 5 shows the UV–vis DRS spectra of 10 wt% Ag/Ni-fiber, 10 wt% Ag/SS-fiber and electrolytic silver catalysts. As noted in previous work,[27-30] the metallic silver particles provide the UV–vis absorption band in the region of 310-320 nm; the isolated Ag^{+} ions and charged $Ag_n^{\delta+}$ clusters deliver those in region of 230-250 nm and 270-290 nm, respectively. As shown in Figure 5, a strong sharp peak at about 320 nm, a weak peak at ~230 nm and a faint peak at ~275 nm, appeared on the electrolytic silver catalyst, indicating the existence of a large amount of metallic silver film or large particles with a relatively low concentration of isolated Ag^{+} ions and charged $Ag_n^{\delta+}$ clusters. In contrast, on the 10 wt% Ag/Ni-fiber and 10 wt% Ag/SS-fiber catalysts, the peak at 320 nm weakened dramatically while the other two peaks especially the one at ~240 nm became strong. This indicates that the 10 wt% Ag/Ni-fiber and 10 wt% Ag/SS-fiber catalysts had higher concentration of isolated Ag^{+} ions and/or $Ag_n^{\delta+}$ clusters than the electrolytic silver catalyst. Not surprisingly, our novel 10 wt% Ag/Ni-fiber and 10 wt% Ag/SS-fiber catalysts provided much higher catalytic activity for the gas-phase selective oxidation of alcohols than the electrolytic silver (Tables 2 to 4, Figures 2 to 4).

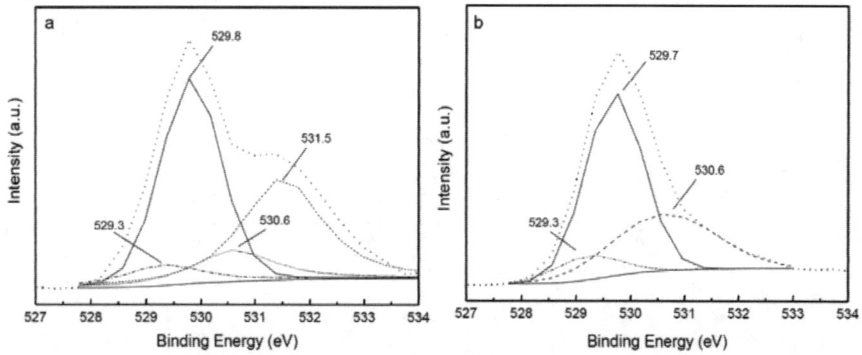

Figure 6. XPS O1*s* spectra of the 10 wt% Ag/Ni-fiber (a) and 10 wt% Ag/SS-fiber (b).

Whereas Ag/Ni-fiber and Ag/SS-fiber catalysts had very similar UV-vis spectra (Figure 5), it seems not safe to say that both of them had equivalent concentration of Ag^{+} ions and $Ag_n^{\delta+}$ clusters. So, XPS measurements for these two catalysts were performed to quantitatively determine the surface concentration of Ag with different chemical states. Nevertheless, it is difficult to identify the chemical state of Ag element from the binding energy (BE) of Ag, because the BE difference between metallic Ag and oxidized Ag is very small and the BE assignment of Ag element remains controversial.[8,31,32] Therefore, the O1s spectra were recorded on the surfaces of the Ag/Ni-fiber and Ag/SS-fiber catalysts as shown in Figure 6. On the surface of Ag/Ni-fiber catalyst, the O existed in four forms with the BE of 529.3, 529.8, 530.6 and 531.5 eV (Figure 6a). The O1s peaks at 529.3 and 530.6 eV were assignable to the strongly chemisorbed O (O_γ) and subsurface O (O_β),[26] which were corresponding to the Ag^{+} ions and $Ag_n^{\delta+}$ clusters; the peaks of O1s at 529.8 and 531.5 eV were associated to the lattice O atoms of NiO and Ni_2O_3, respectively.[32,33] Similarly, on the surface of Ag/SS-fiber, three O1s peaks with BE of 529.3, 529.7 and 530.6 eV were observed (Figure 6b), being ascribable to the strongly chemisorbed O (O_γ) and the lattice O atoms of Fe_2O_3 and Cr_2O_3, respectively.[34-36] For the Ag/SS-fiber sample, the assignment of the O1s peak with BE of 530.6 eV to Cr_2O_3 was strongly confirmed by the facts that the atom ratio of such O to Cr was very close to 1.5 and no clear H_2-reduction peak of O_β species could be detected by H_2-TPR analysis (see posterior section). Moreover, the surface Ag fraction in the total surface metal atoms were determined by XPS to be 25% (*i.e.*, Ag/Ni = 1/3 for 10 wt% Ag/Ni-fiber; Ag/Fe/Cr = 1/2/1 for 10 wt% Ag/SS-fiber), being well consistent with the O_2-chemisorption results (see Table 5). For comparison, the amount of the surface O atoms with the BE of 529.3 eV was normalized by the amount of the surface Ag atoms to indicate the concentration of the Ag^+ ions and $Ag_n^{\delta+}$ clusters in the microfibrous-structure supported silver catalyst samples. Such value number was 0.08 for the 10 wt% Ag/SS-fiber, about a half that (0.14) for the 10 wt% Ag/Ni-fiber. It seems correlated well with the observation that the latter catalyst always provided higher activity than the former one under compatible conditions (see Tables 2 to 4). However, it is not clear yet why the Ag/Ni-fiber delivered much better low-temperature activity/selectivity than the Ag/SS-fiber. For instance, the Ag/Ni-fiber provided a reduction of the reaction temperature by 50-70 °C and delivered compatible conversion with equivalent or higher target product selectivity for the gas-phase selective oxidation of alcohols, in comparison with the Ag/SS-fiber.

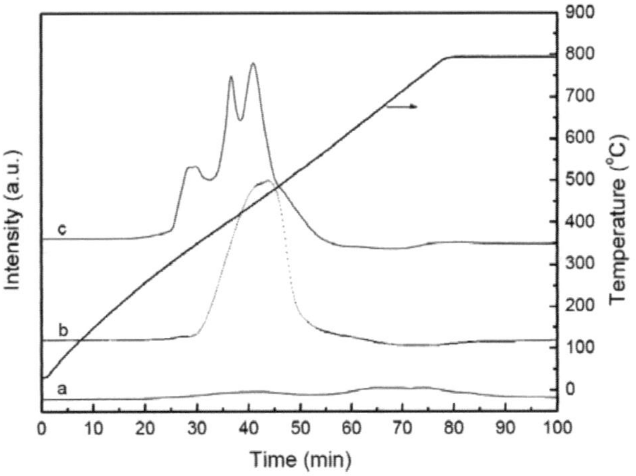

Figure 7. H_2-TPR profiles of electrolytic silver (a, after calcination in air for 2 h), Ni-fiber (b, calcined at 600 °C in air for 4 h), and 10 wt% Ag/Ni-fiber (c).

H_2-TPR and O_2-TPO: catalyst reducibility and ability for O_2 activation

Experiments on the H_2-TPR for the microfibrous-structure supported silver catalysts and electrolytic silver were performed to study their discrepancy in the reducibility with the results as shown in Figure 7. For reference, the microfibrous-structured Ni-fibers and SS-fibers, which experienced 4-h calcination in air at 600 °C, were also studied by H_2-TPR and their profiles were collected in Figure 7. Clearly, the silver catalyst samples studied here presented quite different reduction features. Figure 7a shows the H_2-TPR profile for the electrolytic silver recorded after exposure to O_2 for 2 h at 600 °C. Two overlapped peaks appeared on the profile with the maximum at 450 °C and 530 °C, likely being contributed to the Ag species with subsurface oxygen (O_β) and strongly chemisorbed oxygen (O_γ).[26] Nevertheless, the reduction peak for the Ag species with weakly chemisorbed oxygen (O_α) was not observed clearly. It has been reported that volume diffusion of oxygen into silver is facilitated at elevated temperature thereby leading to the disappearance of O_α.[26] The calcined Ni-fiber sample gave a single-peak profile with the maximum at 450 °C, being ascribable to the NiO (Figure 7b). The 10 wt% Ag/Ni-fiber catalyst offered a profile mainly consisting of an intense peak with a maximum at 430 °C, a strong sharp peak at 400 °C, a weak broad peak at 330 °C and a visible shoulder peak at 500 °C (Figure 7c). According to Figure 7b and the fact that large amount of surface O for NiO and small amount of surface O for Ni_2O_3 were detected by XPS on the 10 wt% Ag/Ni-fiber, the intense peak at 430 °C and the shoulder peak at 500 °C in the profile of Figure 7c were likely assignable to the NiO and Ni_2O_3, respectively. Therefore, the peaks at 400 °C and 330 °C were both contributed to the reduction of Ag species with O_γ and O_β. Actually, both O_γ (O1s: 529.3 eV) and O_β (O1s: 530.6 eV) species were detected by XPS (Figure 6a). Moreover, the H_2 consumption amount from the reduction of Ag species with O_γ and O_β was much larger for the 10 wt% Ag/Ni-fiber sample than those for the electrolytic silver (Figure 7). By comparison, it is safe to say that the 10 wt% Ag/Ni-fiber catalyst had much larger amount of Ag^+ ions and $Ag_n^{\delta+}$ clusters and such Ag sites were more active (corresponding to good low-temperature reducibility), in comparison with the electrolytic silver.

In addition, the 10wt% Ag/SS-fiber and the SS-316L fiber pre-calcined in air at 600 °C for 4 h were also studied by H_2-TPR. Unfortunately, their profiles were complex and similar, having the dominant peaks above 500 °C. By comparison, only a very weak and broad new peak at 480 °C was gleamingly presented for the 10wt% Ag/SS-fiber, likely being assignable to the Ag species with O_γ since such O species were clearly observed on the O1s XPS spectrum for the 10wt% Ag/SS-fiber (Figure 6b). Especially, no clear Ag^+ and $Ag_n^{\delta+}$ reduction peaks compatible to those for the 10wt% Ag/Ni-fiber could be found out below 450 °C on the 10 wt% Ag/SS-fiber. The above results suggest that the 10wt% Ag/SS-316L catalyst had lower concentration of Ag^+ and $Ag_n^{\delta+}$ and such Ag sites were not so active (required higher temperature for H_2-reduction) as those on 10wt% Ag/Ni-fiber. Therefore, like the electrolytic silver, this catalyst showed high activity only at higher temperatures (*e.g.*, 450 °C for the gas-phase selective oxidation of benzyl alcohol).

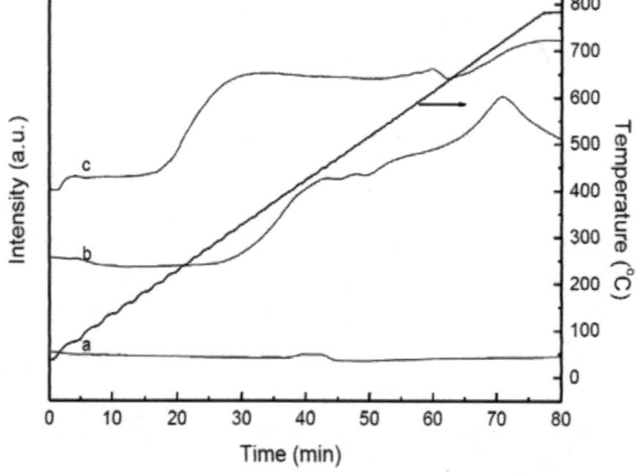

Figure 8. O_2-TPO profiles of electrolytic silver (a), Ni-fiber (b), and 10 wt% Ag/Ni-fiber (c). *Note*: the samples are the corresponding ones after H_2-TPR experiments in Figure 7.

Experiments on the O_2-TPO for the samples after the H_2-TPR tests were also performed for comparative study of their ability to activate O_2, with the results as shown in Figure 8. Figure 8a shows the O_2-TPO profile for the electrolytic silver, giving a very weak O_2 consumption peak with maximum at 420 °C. Ni-fiber sample provided a profile containing multi-overlapped peaks for the oxidation of Ni with the first one centered at ~420 °C (Figure 8b). A profile containing multi-overlapped peaks was also obtained for the 10wt% Ag/Ni-fiber, but a new strong peak appeared with the maximum at 320 °C (Figure 8c) compared with the profile of Ni-fiber sample (Figure 8b). This is likely assignable to the oxidation of the reduced Ag active sites. Such low oxidation temperature indicates that this microfibrous catalyst was much more active for O_2 activation than the electrolytic silver. In addition, O_2 consumption amount concerning the oxidation of reduced Ag sites was much larger for the 10wt% Ag/Ni-fiber than that for the electrolytic silver. This confirms again that the 10wt% Ag/Ni-fiber catalyst had much higher concentration of the active Ag sites, being consistent with the information provided by the H_2-TPR experiments and XPS analyses.

Accordingly, the Ag/Ni-fiber catalyst not only provided high concentration of Ag^+ ions and $Ag_n^{\delta+}$ clusters but also exhibited significantly enhanced low-temperature reducibility and the ability for low-temperature activation of O_2. It is reasonable to explain why the Ag/Ni-fiber catalyst exhibited much better low-temperature activity than the Ag/SS-316L and electrolytic silver catalysts.

Conclusion

A promising thin-sheet microfibrous structured silver catalyst system for the gas-phase selective oxidation of various alcohols was developed by placing Ag onto the surface of the sinter-locked three-dimensional networks consisting of 5vol% 8-μm-Ni- and 12-μm-(SS-316L)-fibers through the incipient wetness impregnation method. The best microfibrous structured silver catalyst was the Ag/Ni-fiber that was calcined at 600 °C in air and had Ag-loading of 10wt% for mono-alcohols or 20wt% for di-alcohols. In comparison with the Ag/SS-fiber and electrolytic silver, the Ag/Ni-fiber catalyst not only provided high concentration of Ag^+ ions and $Ag_n^{\delta+}$ clusters but also exhibited significantly enhanced low-temperature reducibility and the ability for low-temperature activation of O_2 molecules. As a result, such catalyst demonstrated pleasing low-temperature activity/selectivity for the gas-phase selective oxidation of alcohols. For instance, the conversion and product selectivity results on the Ag/Ni-fiber against the electrolytic silver were: 99% and 83% at 330 °C vs. 90% and 42% at 450 °C in the gas-phase selective oxidation of 1,2-propylene glycol to methyl glyoxal; 92% and 86% at 380 °C vs. 55% and 79% at 500 °C in the gas-phase selective oxidation of benzyl alcohol to benzaldehyde. In addition, the microfibrous Ag/Ni-fiber catalyst provided a significant (four-fold or more) promotion of the steady-state volumetric reaction rate compared to the electrolytic silver.

Acknowledgments

Y. Lu gratefully thanks the Program for New Century Excellent Talents in University (NCET-06-0423), Shuguang Project (06SG28), and Shanghai Leading Academic Discipline Project (B409). This work is supported by the National Natural Science Foundation of China (20590366, 20573036), the Ministry of Science and Technology of China (2007AA05Z101).

References

1. Hudlicky M. *Oxidation in organic chemistry*, American Chemical Society: Washington, DC, 1990.
2. Sheldon RA, Kochi JK. *Metal-catalyzed oxidations of organic compounds*, Academic Press: New York, 1981.
3. Mallat T, Baiker A. Oxidation of alcohols with molecular oxygen on solid catalysts. *Chem Rev*. 2004; 104: 3037-3058.
4. Cainelli G, Cardillo G. *Chromium oxidants in organic chemistry*, Springer: Berlin, 1984.
5. Su FZ, Cao Y. Ga-Al mixed-oxide-supported gold nanoparticles with enhanced activity for aerobic alcohol oxidation. *Angew Chem Int Ed*. 2008; 47: 334-337.
6. Sheldon RA, Arends IWCE, Brink GJT, Dijksman A. Green, catalytic oxidations of alcohols. *Acc Chem Res*. 2002; 35: 774-781.
7. Deshpande SS, Jayaram RV. A facile deprotection of oximes over mixed metal oxides under solvent-free conditions. *Catal Commun*. 2008; 9: 639-644.
8. Shen J, Shan W, Zhang Y, Du J, Xu H, Fan K, Shen W, Tang Y. Gas-phase selective oxidation of alcohols: In situ electrolytic nano-silver/zeolite film/copper grid catalyst. *J Catal*. 2006; 237: 94-101.
9. Yamamoto R, Sawayama Y, Shibahara H, Ichihashi Y, Nishiyama S, Tsuruya S. Promoted partial oxidation activity of supported Ag catalysts in the gas-phase catalytic oxidation of benzyl alcohol. *J Catal*. 2005; 234: 308-317.
10. Mao JP, Deng MM, Xue QS, Chen L, Lu Y. Thin-sheet Ag/Ni-fiber catalyst for gas-phase selective oxidation of benzyl alcohol with molecular oxygen, *Catal Commun*. 2009; 10: 1376-1379.
11. Pestryakov AN, Lunin VV, Bogdanchikova NE, Petranovskii VP, Knop-Gericke A. Supported foam-silver catalysts for alcohol partial oxidation. *Catal Commun*. 2003; 4: 327-331.
12. Pestryakov AN, Bogdanchikova NE, Knop-Gericke A. Alcohol selective oxidation over modified foam-silver catalysts. *Catal Today* 2004; 91: 49-52.
13. Magaev OV, Knyazev AS, Vodyankina OV, Dorofeeva NV, Salanov AN, Boronlin AI. Active surface formation and catalytic activity of phosphorus-promoted electrolytic silver in the selective oxidation of ethylene glycol to glyoxal. *Appl Catal* A 2008; 344: 142-149.
14. Pina CD, Falletta E, Rossi M. Highly selective oxidation of benzyl alcohol to benzaldehyde catalyzed by bimetallic gold-copper catalyst, *J Catal*. 2008; 260: 384-386
15. Chang BK, Lu Y, Tatarchuk BJ. Microfibrous entrapment of small catalyst or sorbent particulates for high contacting-efficiency removal of trace contaminants including CO and H_2S from practicle reformates for PEM H_2-O_2 fuel cells. *Chem Eng J*. 2006; 115: 195-202.

16. Lu Y, Sathitsukasnoh N, Queen A, Tatarchuk BJ. Microfibrous Entrapped ZnO/Support Sorbents for High Contacting Efficiency H_2S Removal from Reformate Streams in PEMFC Applications. In: Wang Y, Holladay JD, editors. *Microreactor Technology and Process Intensification.* New York, NY: American Chemical Society Publications Division, Distributed by Oxford University Press, 2005: Chap. 25.
17. Lu Y, Wang H, Liu Y, Xue QS, Chen L, He MY. Novel microfibrous composite bed reactor: high efficiency H_2 production from NH_3 with potential for portable fuel cell power supplies, *Lab Chip* 2007; 7(1): 133-140.
18. Liu Y, Wang H, Li JF, Lu Y, Xue QS, Chen L. Microfibrous entrapped Ni/Al_2O_3 using SS-316 fibers for H_2 production from NH_3, *AIChE J.* 2007; 53: 1845-1849.
19. Wang MM, Li JF, Chen L, Lu Y. Miniature NH_3 cracker based on microfibrous entrapped $Ni\text{-}CeO_2/Al_2O_3$ catalyst monolith for portable fuel cell power supplies, *Intl J Hydrogen Energy* 2009; 34: 1710-1716.
20. Liu Y, Wang H, Li JF, Lu Y, Wu HH, Xue QS, Chen L. Monolithic microfibrous nickel catalyst co-modified with ceria and alumina for miniature hydrogen production via ammonia decomposition. *Appl Catal* A 2007; 328: 77-82.
21. Harris DK, Cahela DR, Tatarchuk BJ. Wet layup and sintering of metal-containing microfibrous composites for chemical processing opportunities. *Compos A: Appl Sci Manuf.* 2001; 32: 1117-1126.
22. Gavriilidis A, Sinno B, Varma A. Influence of loading on metal surface area for $Ag/\alpha\text{-}Al_2O_3$ catalysts. *J Catal.* 1993; 139: 41-47.
23. Pestryakov AN. Modification of silver catalysts for oxidation of methanol to formaldehyde. *Catal Today* 1996; 28: 239-244.
24. Pestryakov AN, Davydov AA. Active electronic states of silver catalysts for methanol selective oxidation. *Appl Catal* A 1994; 120: 7-15.
25. Waterhouse GIN, Bowmaker GA, Metson JB. Oxygen chemisorptions on an electrolytic silver catalyst: a combined TPD and Raman spectroscopic study. *Appl Surf Sci.* 2003; 214: 36-51.
26. Waterhouse GIN, Bowmaker GA, Metson JB. Mechanism and active sites for the partial oxidation of methanol to formaldehyde over an electrolytic silver catalyst. *Appl Catal* A 2004; 265: 85-101.
27. Rhodes HE, Wang P-K, Stokes HT, Slichter CP, Sinfelt JH. NMR of platinum catalysts. I. Line shapes. *Phys Rev* B 1982; 26: 3559-3568.
28. Yu I, Gibson AA, Hunt RE, Halperin WP. Observation of conduction-electron density oscillations at the surface of platinum particles. *Phys Rev Lett.* 1980; 44: 348-351.
29. Stokes HT, Rhodes HE, Wang P-K, Slichter CP, Sinfelt JH. NMR of platinum catalysts. III. Microscopic variation of the Knight shifts. *Phys Rev* B 1982; 26: 3575-3581.
30. Jing F, Tong H, Kong L, Wang C. Electroless gold deposition on silicon(100) wafer based on a seed layer of silver. *Appl Phys* A 2005; 80: 597-600.
31. Ye W-C, Ma C-L, Shang W, Chen Y, Wang R, Wang C-M. Effect of sodium dodecylsulfate on improving microstructural properties of electroplated silver-oxygen-tungsten thin films. *Surf Coat Tech.* 2007; 201: 9456-9461.
32. Wagner CD, Zatko DA, Raymond RH. Use of the oxygen KLL Auger lines in identification of surface chemical states by electron spectroscopy for chemical analysis.

Anal Chem. 1980; 52: 1445-1451.

33. Salvati L, Makovsky LE, Stencel JM, Brown FR, Hercules DM. Surface spectroscopic study of tungsten-alumina catalysts using x-ray photoelectron, ion scattering, and Raman spectroscopies. *J Phys Chem.* 1981; 85: 3700-3707.
34. Kumar KV. XPS core level spectra and Auger parameters for some silver compounds. *J Electron Spectrosc Relat Phemon.* 1991; 56: 273-277.
35. Brion D. Photoelectron spectroscopic study of the surface degradation of pyrite (FeS_2), chalcopyrite ($CuFeS_2$), sphalerite (ZnS), and galena (PbS) in air and water. *Appl Surf Sci.* 1980; 5: 133-152.
36. Tsutsumi T, Ikemoto I, Namikawa T, Kuroda H. X-ray photoelectron spectrum of chromium oxide (Cr_2O_5). *Bull Chem Soc Jpn.* 1981; 54: 913-914.

Multi-Scale Structured Compact Reactor System for Fischer-Tropsch Synthesis

Wei Liu, Energy & Environmental, Pacific Northwest National Lab, Richalnd, WA

Syngas conversion such as Fischer-Tropsch (FT) type of synthesis is critical for production of liquid fuels and/or chemicals through the syngas route. However, the conventional F-T process involves complex reactor designs and process flow diagram, and becomes cost-prohibitive for the plant capacity one or two orders of magnitude smaller than the typical petrochemical plant or oil refinery. Step-out catalytic processing technologies are needed to produce chemicals and liquid fuels from diverse and renewable energy and/or hydrocarbon sources that are characteristic of small capacities and unstable supplies. In this presentation, multi-scale reactor design concepts will be described to make a modular-type reactor unit and achieve the above goal. The modular reactor comprises stacking of alternate catalytic reaction and heat exchange layers. In the reaction layer, the catalyst bed is structured of mini-flow channels in a 3-D thermal conduction matrix. As a result, design parameters at the scales of intrinsic catalyst structures, channel reaction, and reactor vessel can be independently controlled to address the mass and heat transfer problems at the respective scale. The proposed design principles are demonstrated by the exceptional performance achieved with the F-T catalytic reaction. At H2/CO feed ratio of 2, ~95% CO conversion and <10% CH4 selectivity is obtained with the proposed structured bed. By contrast, about 24% CO conversion and 8 % CH4 selectivity is shown with the same catalyst in the crushed particle form. The results suggest it is possible to develop a simple, one-pass, and high-throughput modular-type reactor unit for gas-to-liquid conversion processes .

See more of this Session: Distributed Chemical Processing Technologies
See more of this Group/Topical: Topical G: Innovations of Green Process Engineering for Sustainable Energy and Environment

Production of Autoclaved Cellular Concrete with Aluminum Salt Cake Residues

Xuan Li[1,2], Jiann-Yang Hwang[2]*, Hee-Joon Jeon[2] and Zheng Zhang[2]

[1]Department of Civil and Environmental Engineering, University of Science and Technology Beijing, Beijing, China
[2]Department of Materials Science and Engineering, Michigan Technological University, Houghton, MI
*Corresponding Author: jhwang@mtu.edu

Abstract

Aluminum is recycled through a smelting process. Salt of NaCl and KCl mixture was added during the smelting to seal the air from the aluminum melt and to help dissolving the surface aluminum oxide. After aluminum smelting, the salt cake is disposed. In this study, aluminum salt cake is processed to recover the salt and coarse aluminum particles. The residue powder is rich in aluminum oxide, but still containing a few percent of fine aluminum metal locked in the powder. Attempt is made to substitute the aluminum metal powder in the autoclaved cellular concrete production with the aluminum salt residue. Aluminum can react with water at high pH to generate hydrogen gas. This property has been utilized for the production of autoclaved cellular concrete, where aluminum powder is added to a mixture of cement, lime, water and sand to develop a foamed lightweight concrete structure. This material is lightweight and has good insulation property. However, the use of expensive aluminum powder has limited its broad applications. Results of this study indicated that this aluminum salt cake residue can generated a sufficient amount of H_2 gas, and the product exhibits 50 pcf in density and 400 psi in strength. This research also indicates that the most efficient value of cement addition is 20 wt% for the autoclave processing.

Key words: Aluminum slag ,Autoclaved Cellular Concrete, Fly ash, Concrete

Introduction

Aluminum, the third most abundant element on the earth, is primarily produced from the bauxite ore [1, 2]. Dross is a primary by-product during the smelting process. Significant amounts of metallic aluminum still exist in the dross. To recover the entrained aluminum metal, as well as the aluminum scraps, these materials are repeatedly melted with the additions of salt [3-6]. The salt is generally composed of about 50% NaCl and 50% KCl. At this composition, the salt has a eutectic melting temperature of 657°C, very close to the melting temperature of aluminum (660°C). Salt can dissolve the oxidized surface layer of aluminum while provide a barrier for the oxygen to retard the oxidation of aluminum melt.

The final processing waste is salt cake, which contains aluminum metal that is uneconomical to recover [7]. In general, salt cake consists of 3-5wt% aluminum metal, 20-50wt% salt fluxes, and approximately 45-77wt% non-metallic products (oxides) [8, 9]. After the dissolution of salt fluxes, the aluminum metal content of salt cake increases to 6-8wt%.

Autoclaved Cellular Concrete (ACC) was discovered in Sweden in the early 20th century when a mixture of cement, lime, aluminum powder, water and sand developed a foamed lightweight concrete structure [10]. Generally ACC has 50-75% pore volume in its structure. This substance provides outstanding properties for thermal and acoustic insulation, resulting in savings on heating and cooling costs. Its weight is about 25-33% of the weight of conventional concrete products. It is noncombustible and offers a higher rating of fire resistance. ACC is an inorganic material and does not rot or deteriorate over time. Contribution to environmental concerns is significant by reusing industrial wastes and saving on disposal costs. Even though it is used widely in Europe and has many points of merit, ACC has been slow to gain acceptance in the United States due to the dominant use of wood products in construction.

There are many material factors to consider in the production of ACC. Cement provides the primary binding compound that yields a strength-generating hydration product. Sand is added in order to provide the main source of silica (SiO_2) for a calcium silicate hydrate (C-S-H) product. Cleaned fly ash is used to provide a portion of reactive silicates. Calcium oxide / calcium hydroxide reacts with aluminum metal to generate H_2 gas. This H_2 gas is entrapped by the cementitious mixture and creates numerous gas voids in the concrete. Gypsum slurry, made mainly from calcium sulfate dihydrate, is added to the mixture to keep the gas from escaping the mixture.

Aluminum metal is a primary component of ACC due to a porous structure is developed as a result of the hydroxide-metallic aluminum gas forming reaction. However, the use of expensive aluminum powder has limited ACC broad applications. The salt cake contain large precetage of aluminum metal after desalted, so the generation of H_2 gas takes place when alkali solution and aluminum salt cake are mixed. The objiect of this research is use aluminum slat cake residues to replace metallic aluminum powder to make Autoclaved Cellular Concrete.

Experiment

Materials

Fly ash from Pepco was used in this research. This fly ash was cleaned in a MTU pilot plant froth flotation process to remove the carbon and soluble salts. Density of fly ash is measured as 2.25 g/cc. Standard sand from Superior Sand and Gravel, which has coarse (C) and fine (F) fractions, was used as a main souce of silica. Aluminum slag from a aluminum recycling company was processed for this study. The process includes wet ball mill grinding, screened to -100 mesh, and wash filtered to remove the salts. This slag containing 8.37wt% metallic aluminum, 87.94wt% oxides and 3.69wt% salt. Type I ordinary portland cement made by Lafarge used to strength developing material. Gypsum slurry was prepared by blending 100g gypsum ($CaSO_4 \cdot 2H_2O$), 200g Deionized Water and 0.5g sodium tripolyphosphate ($Na_5P_3O_{10}$). The purpose of the addition of gypsum slurry was to increase H_2 gas holding potential in a cementitious mixture.

Equipment

Cenco-Menzel Autoclave manufacturered by Central Scientific Co. USA was used in this research. Two-inch cube mold containing three compartments was used. Instron 4210 Testing machine was used for measuring strengths of specimens.

Preparation of Specimens

Solids (cement, fly ash and sand) were mixed in a mixing bowl. The comopnents of solids are summarized in Table 1. Deionized water (DW) and CaO were blended at highest speed in a blender for 60 seconds, then add aluminum slag and keep blending until H_2 gas generation occurred. Decanting this slurry into mixing bowl immediately, add gypsum slurry and mixed again until the mixture was uniform. Then the mixture was placed into 2-inch (50mm) cube molds and exposed to ambient conditions for 24 hours. During this period, specimens gained sufficient strength to withstand autoclaving pressure. Another batch of sample was immersed in the lime solution for curing. The curing time was 7 days and 21 days.

Table 1 Components of solids mixture

Sample #	Cement, wt%	Fly Ash, wt%	Sand, wt%	
			c	f
1	10	40	25	25
2	20	20	30	30
3	20	40	20	20
4	20	60	10	10
5	30	40	15	15
6	50	40	5	5

Autoclaving Process

Specimens were removed from the mold and placed into the autoclave equipment(Cenco-Menzel Autoclave) to carried out for 6 hours autoclaving process at 420 °F and 300 psi. After autoclaving, specimens were cooled at ambient conditions for 6-12 hours.

Physical tests

The compressive strengths and density of samples were measured. The density was determined from the specimen weight and measured volume.

Results and Discussion

The blending procedure removes the oxidation layer of aluminum metal in the salt cake residues, leaving exposed metallic aluminum that is reactive to the solution and produces soluble alkali-metal aluminate and hydrogen gas, as shown by equation (1) below:

$$2Al + 2OH^{-} + 6H_2O - 2[Al(OH)_4]^{-} + 3H_2(gas)\uparrow \qquad (1)$$

As a result, it is possible to obtain specimens with lower density than water when H_2 gas is entrapped in the cementitious mixture. The density of final product is directly related to the density of solid particles (such as cement, fly ash, and sand) and gas entrapment. Fly ash is a major component, and its influence on density is more than any other solid raw material.

Calcium silicate hydrate (C-S-H) phase denotes amorphous or nearly amorphous calcium silicate hydrate products of the general formula $CaO_x \cdot SiO_2 \cdot H_2O_y$ where both x and y vary over a wide range. This phase is produced in the hydration of the tricalcium silicate (C_3S) component in a cement. A similar hydration phase is observed in the pozzolanic reaction of fly ash [11-13] and blast furnace slag. C-S-H is a primary hydration product for strength development.

Material strength has proven to be proportional to the amount of C-S-H product. Calcium originates from cement and fly ash, while the silicates are from fly ash and sand.

Autoclaving Process (AC) Effect

To observe a hydration reaction product, X-ray diffractograms of a mixture of solids (sample 2) and deionized water were obtained and illustrated in Figure 1. The compositions of the spectra are 3-hour AC specimen, 6-hour AC specimen and a specimen cured for 7 days in the saturated lime solution. In the diffractogram of 3-hour AC, only the quartz peaks were displayed. In the 6-hour AC XRD spectrum, the intensity of main quartz peak, approximately at 26.5° 2Θ, was reduced and traceable amounts of calcium silicate hydrate start being illustrated at 29.36, 32.24 and 50.08° 2Θ. The most recognized phase was in the form of $Ca_{1.5}SiO_{3.5} \cdot xH_2O$. From the specimen cured for 7-day, similar reduction of quartz main peaks was detected and obvious $Ca_{1.5}SiO_{3.5}\cdot xH_2O$ hydration product was portrayed in the diffractograms.

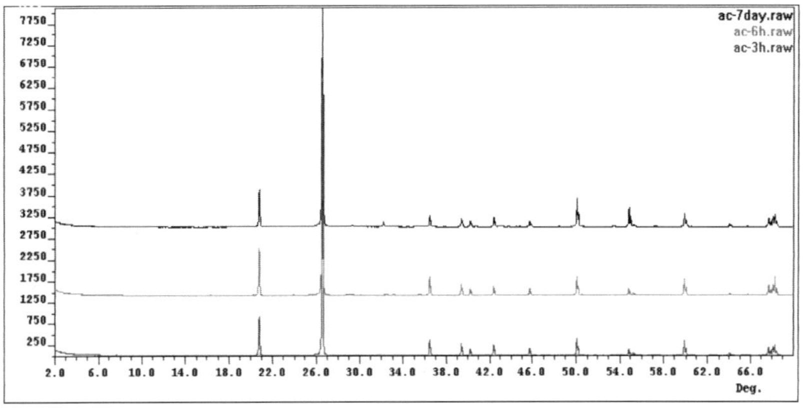

Figure 1. X-Ray diffractograms of AC Processed

To compare specimen physical properties between the autoclaving process and the curing process in saturated lime solution, specimens made by solids sample 2, 3 and 4 were measured. Each sample was added with 4wt% aluminum slag, 8wt% CaO, 5wt% gypsum slurry and 40wt% DW. Figure 2 and 3 display the results. As the amount of fly ash increased, relative density decreased as illustrated in Figure 2, due obviously to the low specific gravity of fly ash. As the portion of fly ash was increased, the amount of sand was decreased in the mixture. Among solid materials, cement has a 3.15 specific gravity, whereas fly ash and sand have 2.25 and 2.8, respectively.

Figure 2 also shows the density of products from the AC process is higher than that produced from the solution curing process. This is believed due to the more complete CSH polymerization reaction by AC. Strength progress yielded by AC is approximately 2.5 times that of the 28-day curing process in saturated lime solution as portrayed in Figure 3. High strength development from AC is another evidence to support the development of the C-S-H reaction. As the amount of C-S-H production is increased, the density and the strength of the cured cementitious product are increased. The result also show that sample contained 20wt% cement, 40wt% fly ash and 40wt% sand has the most efficient value of strength by autoclaving.

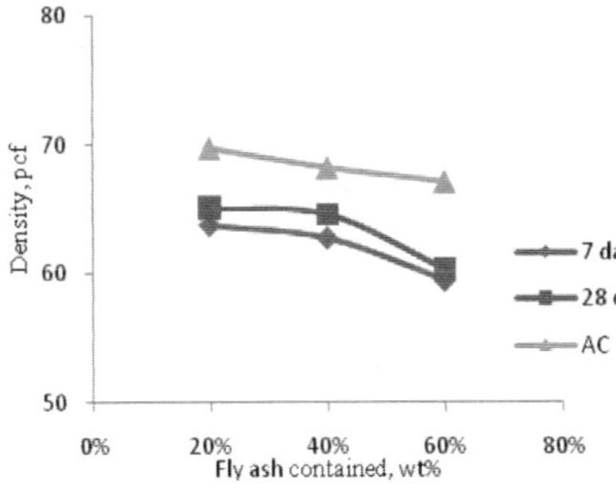

Figure 2. AC effect on density

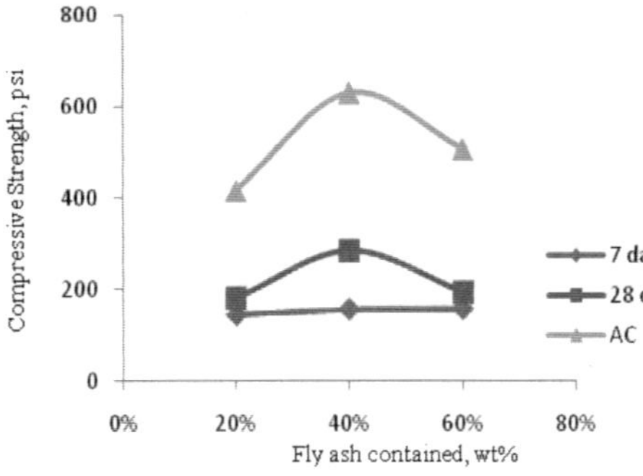

Figure 3. AC effect on Compressive Strength

Cement Effect on AC Process

To observe the cement effect on the curing process, properties of specimens 1, 3, 5 and 6 (4wt% aluminum slag, 8wt% Cao, 5wt% gypsum slurry and 50wt% were added in each sample) were measured. Figures 4 and 5 show the result.

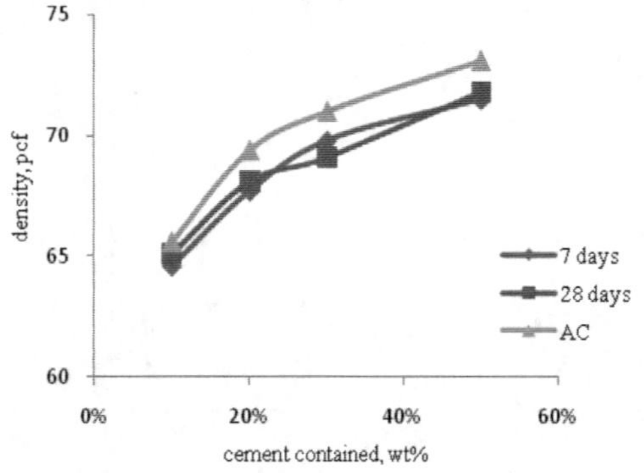

Figure 4. Cement effect on density

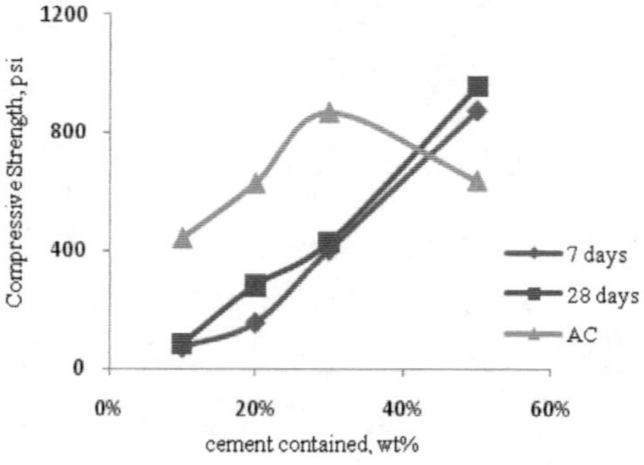

Figure 5. Cement effect on compressive strength

As shown in Figure 4, relative density increases with the increase of amount of cement. This is due to the high specific gravity of cement. The density of concrete by AC process are higher than that by solution curing process.

As shown in Figure 5, the autoclaved samples with 10 to 40wt% cement addition have much higher strengths over the samples cured in lime solutions. The maximum strenth was obtained at 30% cement addition. The mass balance between the lime and slica for C-S-H reaction is probably at optimum at this ratio. Curing in solution have little difference in strength between the 7 and 28 days curing time. This curing means obviously is not effective.

CaO and aluminum slag effect on Autoclaving Process

The effect of CaO content on the autoclaved process was investigated. Specimens with compositions in ratios of sample 1 (with the addition of 4wt% aluminum slag, 5wt% gypsum slurry and 50wt% DW) and variable amount of CaO were measured. Figure 6 show the results.

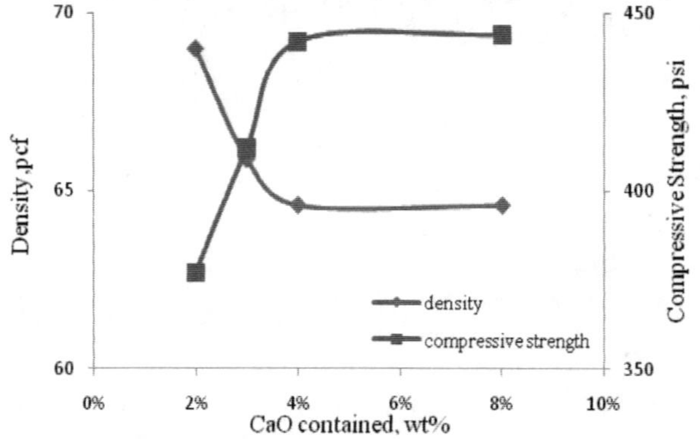

Figure 6. CaO effect on density and strength in AC condition

The effect of aluminum slag content on the autoclaved process was also investigated.

Specimens with compositions in ratios of sample 1 (with the addition of 4wt% aluminum slag, 5wt% gypsum slurry and 50wt% DW) and variable amount of CaO were measured. Figure 8 show the results.

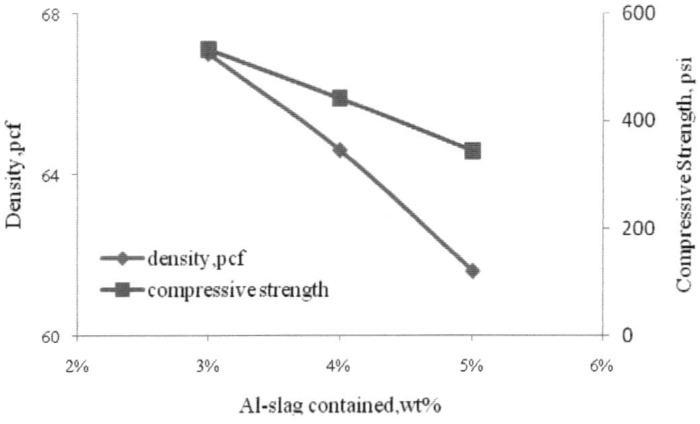

Figure 7. Al-slag effect on density and strength in AC condition

Strength development and high volume of expansion are reciprocal in nature. Cement and CaO contribute to strength, while expansion volume depends on the quantity of reaction products between CaO and aluminum slag.

When the amount of CaO was increased from 2 to 4wt%, significant reduction on density was observed, as portrayed in Figure 6. The reduction of density was believed due to the increase of H_2 gas generation. From 4 to 6wt% CaO, no change on density was observed. Probably the available aluminum is consumed up. Strenth is increased at 2 to 4wt% of CaO, but also becomes flat from 4 to 6wt% CaO, as shown in Figure 6. Increased C-S-H activity should be responsible for the increase.

The increase of the amount of aluminum slag also resulted in density reduction, as shown in Figure 7. The increase in aluminum slag created higher expansion due to more H_2 gas produced. However, without the increase of CaO to yield more C-S-H reactions, the strength decreases quickly with the addition of aluminum slag.

Conclusion

Aluminum slag can be utilized to generate H_2 gas for a porous concrete structure. It has the potential to replace commercially employed metallic aluminum powder for the production of lightweight concrete. Based on various recipes tested, autoclave curing is a better curing approach than the solution curing process to derive low density but high strength products. The C-S-H reaction is believed responsible for the increase of strength.

References

1. W. Bucher, R. Schliebs, G. Winter, K. H. Bücher, "Industrial Inorganic Chemistry" 1989, pp 247-250

2. Aluminum Manufacturing, Pollution Prevention and Abatement Handbook, World Bank Group, July 1998

3. Speaker, Peter G. Schirk. (1997), "Complete Aluminum Dross Processing Technology" in TMS

Annual Meeting, Orlando, Florida.
4. Speaker, D. J. Roth.(1996), “Recovery of Aluminum From Rotary Furnace Salt Cake by Low Impact Rotary Tumbling” in TMS Annual Meeting, Los Angeles California.
5. Speaker, Henrik Gripenberg. (1997), “Salt-Free Dross Processing With ALUREC - Two Years Experience” in TMS Annual Meeting, Orlando, Florida.
6. Dan Yerushalmi, “Aluminum Dross Recovery Process” US Patent 05102453, 1991
7. J. Hwang, “Recycling and Reuse of Aluminum Wastes” (Report DE-FC07-98ID13663, Michigan Technological University, 2000).
8. Speaker, Donald G. Graczyk. (1997), “Analytical Chemistry of Aluminum Salt Cake” in TMS Annual Meeting, Alice M. Essling and Edmund A. Huff, ed. Orlando, Florida.
9. Speaker, D. L. Stewart, Jr.(1997) “Quantifying The Amount of Oxides in Aluminum” in TMS Annual Meeting, K. M. Tomaswick. Orlando, Florida.
10. Autoclaved Aerated Concrete, Environmental Building News, Vol. 5 No.2, March/April 1996
11. Richard Helmuth, “Fly Ash in Cement and Concrete – Lime Pozzolan Reaction Mechanisms”, pp. 99-100
12. Takemoto K. and Uchigawa H., “Hydration of Pozzolanic Cement,” Vol. I, pp. IV-2/ 1-29 in 17B
13. Clifton J. R., Brown P. W. and Frohnsdorff G., “Reactivity of Fly Ashes with Cement,” in Cements Research Progress 1977, Ch. 15, pp. 321-341.

High Efficiency Nitrogen Fertilizer Production From Power Plant Fly Ash

Ben Wen, Advanced Materials, United Environment & Energy LLC, Horseheads, NY

This research concerns a high efficiency, low cost nitrogen fertilizer manufacturing technology from fly ash, which is a recycled material from coal power plants that may contain high concentrations of mercury and carbon. Traditional nitrogen fertilizer production uses natural gas as the primary feedstock and is very costly. The traditional fertilizers are water-soluble compounds, resulting in significant loss of fertilizer and pollution of streams and ground water. At the same time, America's coal power plants produce more than 71.1 million tons of fly ash per year, and most of it is disposed of in landfills. The high mercury content in the fly ash makes the disposal more difficult and costly. The objectives of this project were to develop a high volume and highly technical application for fly ash and bring a value-added high efficiency low cost nitrogen fertilizer product to farmers.

See more of this Session: Conversion of Solid Wastes to Energy and/or Product
See more of this Group/Topical: Topical G: Innovations of Green Process Engineering for Sustainable Energy and Environment

Research Into Design and Modeling of Anaerobic Digestion Process Applied to Municipal Solid Wastes

Sulin Chen, Department of Biological Systems Engineering, Washington State University, Pullman, WA

Usama Zaher, Department of Biological Systems Engineering, Washington State University, Pullman, WA

Tim Ewing, Department of Biological Systems Engineering, Washington State University, Pullman, WA

Paul Gamble, Department of Biological Systems Engineering, Washington State University, Pullman, WA

Anaerobic digestion (AD) potentially offers a much better option for management of the organic fraction of municipal solid wastes compared with landfill and incineration. AD not only harvests the organic carbon in the form of methane as bioenergy, but also makes it easier to recover and recycle the nutrients contained in the wastes. Additionally, AD system can be constructed near the source of the wastes, thus long distance trucking is eliminated resulting in lower transportation costs of the materials which typically has high water contents. Adoption of AD for municipal waste management however, relies on the development of new technologies that offers required reliability and efficiency at lowest possible costs. This presentation summarizes the research progress made at Washington State University in developing AD technologies with high solids contents aimed at addressing these challenges. Main topics cover (1) concept and test results of a new system design, (2) mixing strategy, (3) modeling and simulations of different feedstocks, and (4) comparison with other processes such as composting, landfill and incineration. The information represents some new developments in AD research and development related to municipal solid waste applications.

See more of this Session: Conversion of Solid Wastes to Energy and/or Product
See more of this Group/Topical: Topical G: Innovations of Green Process Engineering for Sustainable Energy and Environment

Municipal Solid Waste (MSW) to Liquid Fuels Synthesis: A Techno-Economic Evaluation of the Production of Mixed Alcohols

Susanne B. Jones, Energy and Environment, Pacific Northwest National Laboratory, Richland, WA

Yunhua Zhu, Energy and Environment Directorate, Pacific Northwest National Laboratory, Richland, WA

Corinne Valkenburg, Energy and Environment, Pacific Northwest National Laboratory, Richland, WA

Municipal solid waste (MSW) is a potential gasifier feedstock that presents opportunity to produce alternative liquid fuels because of its availability in significant amounts at current landfills and because it is a predominantly biomass-derived material that, like conventional biomass feedstocks (such as wood), is a mostly renewable resource. The viability of MSW as a gasifier feedstock for liquid fuels synthesis depends on several factors. Foremost is the availability of MSW in sufficient quantities to meet the minimum process scale required for economic feasibility. That scale is identified, in this study, to be 3,300 short tons per day (as received) or more of MSW. A preliminary review of available information on the number and size of various landfill sites around the country identified 47 sites that meet the minimum scale identified for economic feasibility. Together these sites could potentially produce enough liquid fuel to meet approximately 1.4% of current transportation fuel demand (about 113 MM bbl/year of liquid fuel). A greater contribution could be attained if smaller scale facilities are found to be feasible due to latitude in the tipping fee charged to MSW producers. This work provides a preliminary techno-economic analysis for the production of mixed alcohols from MSW. This study investigates MSW, as it is processed into refuse derived fuel (RDF) and then gasified in a plant co-located with a landfill. The resulting syngas is then catalytically converted to mixed alcohols. At a scale of 2200 short tons per day of RDF, and using current technology, the minimum ethanol selling price at a 10% rate of return is approximately $1.85/gallon ethanol. Favorable economics are dependent upon the toxicity characteristics of the waste streams.

See more of this Session: Conversion of Solid Wastes to Energy and/or Product
See more of this Group/Topical: Topical G: Innovations of Green Process Engineering for Sustainable Energy and Environment

Porous Material from Biomass Process Residue Rich in Cellulose and Its Application

Qiang Li, Yingli Wang, Mei Zhong, Jian Yu, Yin Wang and Guangwen Xu

State Key Laboratory of Multi-Phase Complex System, Institute of Process Engineering, Chinese Academy of Sciences, P. O. Box 353, Beijing 100190, China

1. Introduction

The processes of light industry with agricultural feedstock have to produce solid wastes, which are also called the process residues, along with its conversion of the feedstock into products of food, drink and medicine. These process residues include bagasse, lees of distilled spirit, alcohol and vinegar, grounds of tea and coffee, cakes of oil and soy sauce, and so forth. Among these residues, lees of distilled spirit and vinegar are the residues rich in cellulose and their amount is more than 20 millions of tons every year in China[1] and, most of them are landfilled or discarded in the open air[2-4]. Landfill and open discarding cause surely serious environmental pollution and are waste of biomass resource. Therefore, it is necessary to develop new technology to utilize these biomass lees in high value and large capacity.

The lees of distilled spirit and vinegar contain rice husk up to 50 wt.%. It is reported that highly amorphous silica-base porous material can be manufactured with rice husk[5, 6]. In this study, supposing that the silica in the residues can be a kind of structure frame, we tested the approach of manufacturing porous materials with spirit lees (SL) and vinegar lees (VL), and then examined their applications to aqueous phenol adsorption and NO catalytic reduction.

2. Experimental section

2.1. Sample

The employed spirit lees (Sichuan, China) and vinegar lees (Beijing, China) were dried in an electrical oven at 110℃ for 4h and 2h respectively, before manufacturing porous materials. Their chemical analysis is presented in Table 1.

Table 1. Properties of the spirit lees and vinegar lees

	Proximate analysis db.wt.%			Ultimate analysis daf. wt.%				
	Ash	V.M.	F.C.	C	H	N	O	S
SL	12.7	70.4	16.9	43.80	6.31	3.42	33.00	0.77
VL	11.1	72.6	16.3	41.04	3.99	1.76	41.41	0.66

2.2. Material preparation

The procedure of preparing the porous materials included mainly the steps of carbonization, activation, washing and drying. The activators are CO_2 and steam.

In this study, a commercial granular activated carbon from coconut-shell (CSAC) and coal (CAC) were also used to compare with the newly made porous materials. All of the materials were ground and sieved to sizes below 0.2 mm.

2.3. Adsorption and catalyst tests

Adsorption of phenol from aqueous solution was tested under following condition: The porous material (0.2g) was blended into 20.0 ml phenol solution which concentration was 200mg/l and shaken at 25°C for 12h.

The porous materials from vinegar lees (VL-PM), CSAC and CAC were impregnated using an aqueous solution of ammonium metavanadate in oxalic acid ($m(V_2O_5)/m(\text{support})$ = 1: 99). Thereafter, the catalysts were dried and subsequently calcined at 100℃ and 350℃, respectively. The activity test of the catalysts was conducted with a fixed bed reactor.

3. Results and discussion

3.1. Porous material from vinegar lees

Three kinds of porous materials (named as PM-A, PM-B and PM-C) were prepared with vinegar lees and Table 1 shows the preparing conditions and the corresponding BET specific surface area of the materials.

Table 1. Conditions of preparing the porous materials with vinegar lees and BET specific surface area (S_{BET})

Sample	Carbonization		CO_2 activation		S_{BET}
	T(℃)	time	T (℃)	time	(m^2/g)
PM-A	800	1h	875	1h	482
PM-B	800	1h	875	3h	469
PM-C	800	1h	800	1h	270
CSAC					410

It can be seen that PM-A, PM-B and CSAC had larger BET specific surface area (S_{BET}) than PM-C. The values of S_{BET} for PM-A, PM-B, PM-C and

CSAC were 482, 469, 270 and 410 m^2/g, respectively. These results indicated that better condition for preparing the VL-base porous material was the carbonization at 800℃ for 1.0 h in N_2 atmosphere and then the activation at 875 ℃ for another 1.0 h in CO_2 atmosphere.

The scanning electron micrographs (SEM) of materials PM-A and CSAC (Figure 1) show that the external surface of PM-A was full of small pores, whereas much fewer pores on the surface of CSAC. This difference is possibly caused by the frame composition of the materials. In PM-A, the frame consists of Si and C which form Si-C base, while the CSAC frame is made of C only. Therefore, the material with Si-C frame facilitated the formation of pore structure in thermal treatment process, such as carbonization. In fact, the presence of Si element in the material frame made the volatiles release only from the C sites during carbonization, consequently resulting in high gas release velocities to cause large pores.

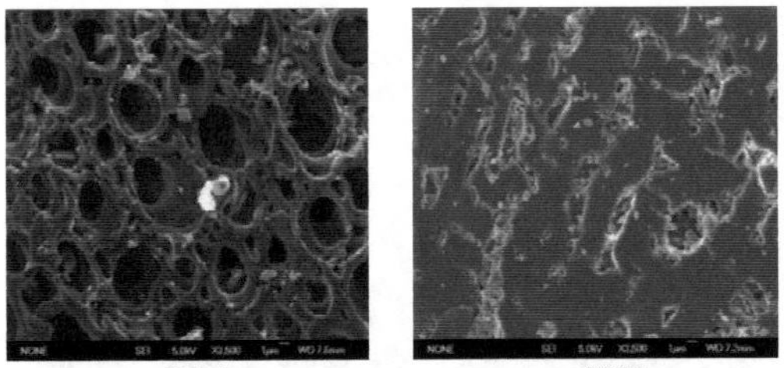

PM-A CSCA

Figure 1. SEM photographs of different materials

3.2. Porous material from spirit lees

The effects of carbonization and activation temperatures on the structure characteristics of the porous materials were investigated and the results were shown in Figures 2 and 3. When CO_2 was used as the activator (Figure 2), the values of iodine (I_2) adsorption and methylene blue (MB) adsorption decreased with the carbonization temperature increasing. Herein, the value of I_2 adsorbed represents the micropore performance, while mesopore performance for that of MB absorbed. This result indicated that lower temperature carbonization was benefit to form porous material with good pore structure. However, the values of I_2 and MB adsorption increased with the activation temperature increasing from 750 to 850℃, while almost not increasing at 900℃. This result suggested that 850℃ was enough for activating the carbonized materials.

When H_2O was used as the activator (Figure 3), similar tendency of the variation of I_2 and MB adsorption values with those shown in Figure 2 was observed. However, it was also observed that under the same temperatures of carbonization and activation, the values of I_2 and MB adsorption for H_2O as the activator were larger than those for CO_2 as the activator. For example, under 450℃ of the carbonization temperature and 850℃ of the activation temperature, the values of I_2 and MB adsorption for H_2O as the activator were 524.3 and 98.7 mg/g respectively, whereas 468.1 and 79.1 mg/g respectively for CO_2 as the activator. This result indicated that the activation of the carbonized material in H_2O was better than that in CO_2 for developing the pore structure of the porous material. Figure 4 also showed that the pore volumes and average pore diameters of the porous materials for H_2O as the activator were larger than those for CO_2 as the activator. For example, at 800℃, the pore volume and average pore diameter of the porous material for H_2O as the activator were 0.32 cm^3/g and 3.52 nm respectively, whereas 0.24 cm^3/g and 3.05 nm respectively for CO_2 as the activator. That is to say, steam can improve the conversion of the pore from micropore to mesopore than CO_2.

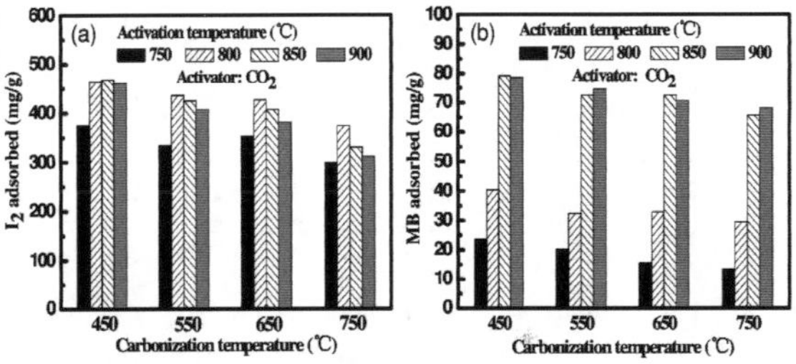

Figure 2. Effects of carbonization and activation temperatures on the structure characteristics of the porous materials (CO_2 as the activator)

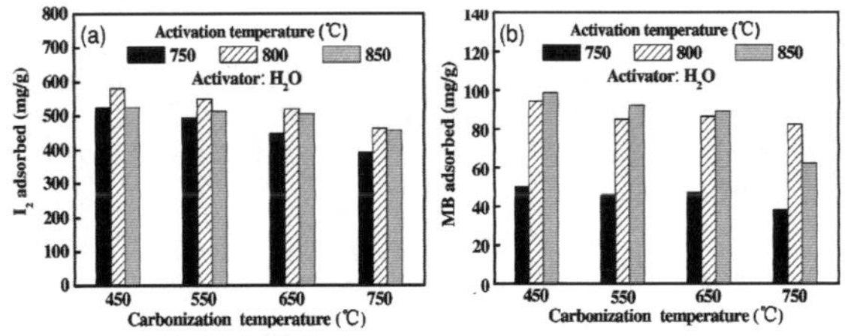

Figure 3. Effects of carbonization and activation temperatures on the structure characteristics of the porous materials (H_2O as the activator)

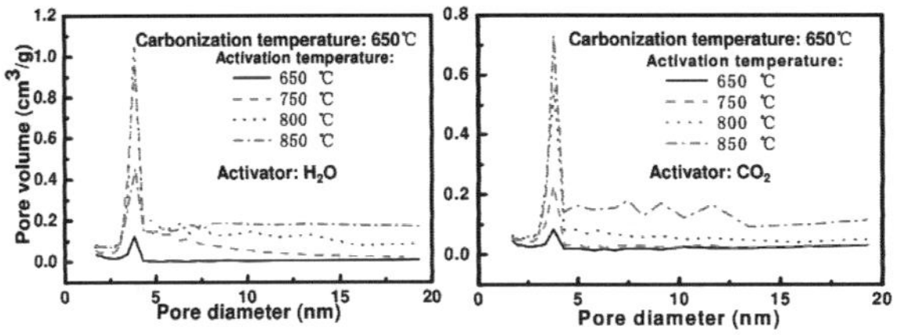

Figure 4. Pore distributions of the porous materials

4. Application

4.1. Aqueous phenol adsorption

Adsorption of phenol was tested by putting 0.2g of porous material from VL into 20 ml of aqueous solution containing phenol of 200 mg/l. Figure 5 compares the realized adsorption for all the porous materials from VL and CSAC, demonstrating that the removed phenol by adsorption was evidently lower for PM-C. This may be caused by the different activation temperatures for PM-A to PM-C. A higher activation temperature possibly led to a larger BET surface area, and then a larger phenol adsorption, as PM-A (875℃) compared with PM-C (800℃). The almost same phenol removal (nearing 100%) by PM-A, PM-B and CSAC proved that it is feasible to manufacture porous material with good adsorption capacity by using VL, a kind of biomass waste.

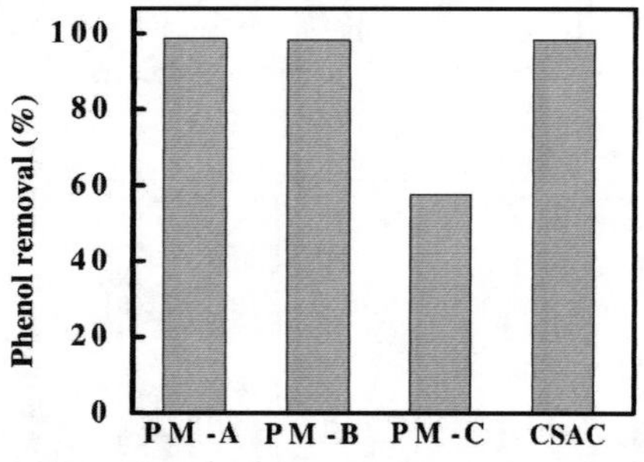

Figure 5. Phenol removal of different porous materials

4.2. Catalyst support

Figure 6 showed the variation of NO conversion over V/VL-PM, V/CSAC and V/CAC with reaction temperature. The NO conversion on all catalysts increased with reaction temperature increasing except for the V/CAC catalyst. The SCR activity order of catalysts was V/VL-PM> V/CSAC> V/CAC. The NO conversion was higher over V/VL-PM than those over V/CSAC and V/CAC, reaching to 60%. This may be caused by the hydrophobicity of activated carbon support for possessing amorphous SiO_2.

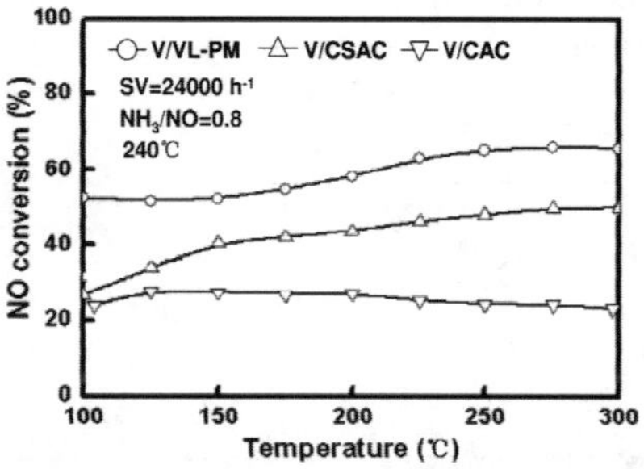

Figure 6. Comparison of the NO conversion over different catalysts with temperature.

4.2 Effect of SO_2

SO_2 is one of the main pollutant components in flue gas and often leads to catalyst deactivation during reducing NO_x. As shown in Figure 7, the activity of V/CSAC decreased under SO_2 introduced into the flue gas, suggesting that the catalyst was deactivated for depositing ammonia sulfate. However, the NO conversion over V/VL-PM increased in the presence of SO_2, indicating VL-base porous material could promote the decomposition of ammonium sulfate. Thus the low temperature activity of vanadium based carbon catalyst in the presence of 300ppm SO_2 was as follow: V/VL-PM>V/CSAC>V/CAC.

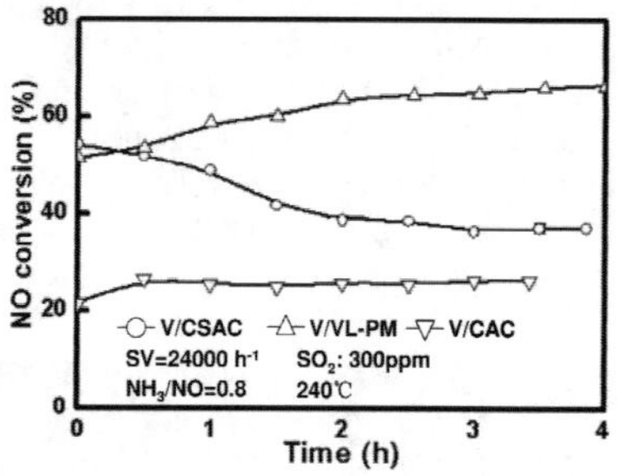

Figure 7. Comparison of the NO conversion over different catalysts in the presence of SO_2

5. Conclusions

Using lees of distilled spirit and vinegar as raw feedstock, porous materials were prepared through carbonation and activation, and then were applied to phenol removal from aqueous solution and catalyst support to reducing NO. The results show that better condition for preparing VL-base porous material was carbonization at 800 ℃ for 1.0 h in N_2 and activation at 875 ℃ for 1.0 h in CO_2. The achieved specific surface area of the material was larger than that of a commercial coconut shell activated carbon. For preparing SL-base porous material, it was concluded that low temperature carbonization and high temperature activation was better experimental condition. The activation of the carbonized material in H_2O was better than that in CO_2 for developing the pore structure of the porous material. Steam can improve the formation of mesopore than CO_2.

Applying the VL-base material to adsorption of aqueous phenol demonstrated that the material realized higher phenol adsorption than CSAC did. It was observed that VL-PM can be used as the catalyst support of low temperature SCR. Compared with other active carbon support, vanadium impregnated on VL-PM had better low temperature activity and SO_2 resistance.

Acknowledgements

The research was supported by the National

High-Tech Research and Development Program of China (863). Acknowledgement would be also extended to the National Natural Science Foundation of China (NSFC) for its support on related fundamental studies (20606034, 20776144).

References

(1). Xu, G. W.; Ji, W. F.; Wan, Y. H.; Liu, C. C.,. Energy production with light industrial biomass residues rich in cellulose. *Progress in Chemistry (in Chinese)*, **2007**,19, 1164-1176.

(2). Ma, X. J.; Chen, J. Y.; Zhang, R. Y.; Han, X. L., Utilization foreground of distiller's grains. *Liquor-making Science and Technology (in Chinese)* **2006**,4, 96-98.

(3). Shi, A. H., Present and future of environmentally friendly distillers' grains in China. *Chinese Brewage (in Chinese)* **2006**, 3, 4-7.

(4). Li, Z. X., Study on development and utilization of vinegar lees germ bran. *Fertilizer Research (in Chinese)* **2000**,8, 31-32.

(5). Kumagai, S.; Sasaki, K.; Shimizu, Y.; Takeda, K., Formaldehyde and acetaldehyde adsorption properties of heat-treated rice husks. *Separation and Purification Technology* **2008,** 61, (3), 398-403.

(6). Chandrasekhar, S.; Satyanarayana, K. G.; Pramada, P. N.; Raghavan, P., Review Processing, properties and applications of reactive silica from rice husk—an overview. *Journal of Materials Science* **2003,** 38, 3159 – 3168.

Pretreatment of Hog Manure Prior to Anaerobic Digestion

Elsayed Elbeshbishy[1], Saad Aldin[2], George Nakhla[1*], Madhumita B. Ray[2]

[1]Dept. of Civil and Environmental Engineering,[2]Dept. of Chemical and Biochemical Engineering, University of Western Ontario, London, Canada, London, Ontario, Canada, N6A 5B9

Abstract

The efficiency of ultrasonication as a pretreatment process of hog manure prior to anaerobic digestion is evaluated in this study. The effect of ultrasonication on particle size distribution, chemical oxygen demand and methane production was studied. The mean particle size diameter decreased from 59 to 21.9 μm with the SSA increasing from 0.523 to 1.2 $\mu m^2\ g^{-1}$ at a specific energy of 30,000 kJ $kg^{-1}\ TS^{-1}$. For specific energy input of 10,000 kJ $kg^{-1}\ TS^{-1}$, a 29.6% reduction in VSS is achieved. The maximum solubilisation of hog manure measured as $COD_{solubilisation}$ was 27.3%. The optimum specific energy input for methane production was 500 kJ $kg^{-1}\ TS^{-1}$, and resulted a 28% increase in methane production, subsequently about $ 4.1 ton^{-1} of dry solids excess energy.

Keywords: Pretreatment; Anaerobic digestion; Ultrasound; Hog manure.

1. Introduction

Ultrasonication has been widely tested to improve the hydrolysis rate in anaerobic digestion of biosolids (Gonze, et al., 2003; Navaneethan, 2007). Ultrasonication disrupts biosolids flocs and bacterial cells, releasing intracellular components, subsequently improving the rate of anaerobic

* Corresponding author. Tel.: +1 519 661 2111x85470; fax: +1 519 850 2991.

E-mail addresses: gnakhla@fes.engga.uwo.ca, gnakhla@eng.uwo.ca (G. Nakhla).

digestion due to the solubilisation of the particulate matter, decreasing SRT and improving the overall performance of anaerobic digestion (Pavlostathis and Gosset, 1986). Tiehm et al., (2001) achieved a reduction in SRT from 20 to 8 d applying ultrasonication on waste activated sludge (WAS) and observed a significant increase in biogas production. The use of ultrasonication in the pretreatment of WAS improved the operational reliability of anaerobic digesters, decreased. However, economical feasibility and durability due to erosion of the sonotrode and high energy inputs are major challenges that need to be resolved for the technology to be adoptable at large scale (Perez-Elvira et al., 2006). Sludge characteristics such as total solids and particle size may highly impact the disintegration efficiency and improve the overall economy of the process. Pretreatment of sludges with high solids content and wide ranges of particle sizes has never been evaluated earlier. In this work, we report the results of ultrasonication of hog manure, which is characterized with high TS and COD content, and with large size distribution.

Therefore, the overall objectives of this study are to evaluate the impact of ultrasonication on anaerobic biodegradability of hog manure by assessing specifically the following: i) the effect of specific energy input as a function of total solids and disintegration efficiency, and ii) biomass-specific gas production rate. Moreover, this paper will include a new approach for the evaluation of disintegration efficiency using soluble and particulate chemical oxygen demand, which are both cost and time effective than the earlier methods reported in literature (Kunz and Wagner, 1994; Muller and Pelletier, 1998).

2. Material and methods

2.1. Analytical methods

Samples were analyzed for total solids (TS), volatile solids (VS), volatile suspended solids (VSS) using standard methods (APHA, 1998). Total and soluble chemical oxygen demand (TCOD, SCOD) were measured using HACH methods and test kits (HACH Odyssey DR/2500). Soluble parameters were determined after filtering the samples through 0.45 μm filter paper. Particle size distribution was determined by Malvern Mastersizer 2000 (version 5.22) laser beam diffraction granulometer.

2.2. Experimental set-up

A lab scale ultrasonic probe was used to treat hog manure obtained from local hog farm in Southwestern, Ontario, Canada. The characteristics of hog manure in mg L^{-1} used in this study as following: TCOD (144900), SCOD (55800), TS (93180) and VS (66980). The ultrasonic probe was supplied by Sonic and Materials (model VC-500, 500 W, and 20 kHz). Hog manure was sonicated for different specific energy inputs, with sonication pulses set to 2 seconds on and 2 seconds off to control the temperature rise of the sludge.

2.3. Specific energy input:

The specific energy input (SE) is a function of ultrasonic power, ultrasonic duration, and volume of sonicated sludge and TS concentration, and can be calculated using the following equation Bougrier et al., (2005):

$$SE = \frac{P \times t}{V \times TS} \qquad (1)$$

where *SE* is the specific energy input in kWs kg^{-1} TS^{-1} (kJ kg^{-1} TS^{-1}), P the ultrasonic power in kW, t the ultrasonic duration in seconds, V the volume of sonicated sludge in litres, and TS the total solids concentration in kg L^{-1}.

2.4. Degree of disintegration (DD):

In this study, the degree of disintegration was determined based on the equation of Muller and Pelletier, (1998) as defined below;

$$DD_M = \left[\frac{COD_{Ultrsound} - COD_{Original}}{COD_{NaOH} - COD_{Original}} \right] \times 100\% \qquad (2)$$

where $COD_{ultrasound}$ is the COD of supernatant of ultrasound treated sample (mg L^{-1}), $COD_{original}$ is the COD of supernatant of original (untreated) sample (mg L^{-1}), COD_{NaOH} is the maximum COD release in the supernatant after NaOH digestion. The NaOH digestion is carried out by treating the sludge samples with 1 M NaOH at a ratio of 1:2 by volume for 10 min at 90°C.

2.5. $COD_{solubilisation}$

$COD_{solubilisation}$ was calculated using the SCOD released which is the difference between SCOD at any time after ultrasonication ($SCOD_t$) and the initial SCOD ($SCOD_0$) divided by the initial particulate COD ($TCOD_i - SCOD_0$):

$$\mathrm{COD}_{\mathrm{solubilisation}} = \left[\frac{SCOD_t - SCOD_0}{TCOD_i - SCOD_0} \right] \times 100\% \qquad (4)$$

Where $TCOD_i$ is initial TCOD.

3. Results and Discussion

3.1. Particle size distribution

Particle size distribution is widely used as qualitative measure for sludge disintegration. Anaerobic digestion of waste is governed by hydrolysis (solubilisation of particulates) that is highly affected by the particle size. Smaller particle sizes and the lower concentration of particulates, measured as VSS lead to higher degradation efficiency. As shown in Fig. 1, the hog manure is characterized by a wide range of particle size ranging from 0.6 to 2,500 μm, compared to a range of 0.4 to 1,000 μm reported for WAS (Gonze et al., 2003; Bougrier et al., 2005). As shown in Fig. 1, the particle size distribution for the hog manure shows a bi-modal distribution, with two peaks, first at 60 and another at 1,200 μm, respectively. Interestingly, the disintegration effect was more pronounced for the particles in the range of 0.6 to 60 μm; while a minor effect was observed for particles > 60 μm. The mean particle size diameter decreased from 59 to 21.9 μm with the SSA increasing from 0.523 to 1.2 $\mu m^2\ g^{-1}$ at a specific energy of 30,000 kJ $kg^{-1}\ TS^{-1}$ (Table 1). Using WAS, Tiehm et al., (2001); Gonze et al., (2003); Bougrier et al., (2005) achieved minimum mean diameters of 20, 18.1 and 12.7μm, respectively. On the other hand the average reduction in VS was 22.5 ± 2% for the specific energy in the range of 500 to 5,000 kJ $kg^{-1}\ TS^{-1}$. For hog manure used in this work, increasing the specific energy to 10,000 kJ $kg^{-1}\ TS^{-1}$ raised the VS reduction percentage to 29.6%, while increasing the specific energy beyond 10,000 kJ $kg^{-1}\ TS^{-1}$ did not improve the VS reduction significantly.

3.2. $COD_{solubilisation}$

Ultrasonic pretreatment solubilises extracellular matter and extracellular polymeric substances (ESP), increasing the SCOD. Thus, SCOD is mostly used to measure the sludge disintegration efficiency that evaluates the degree of solubilisation of sludges. The specific energies for various TS% and DD from this study and two other studies are plotted in Fig. 2. A sharp decline in the required specific energy from 65,000 to 10,000 kJ $kg^{-1}\ TS^{-1}$ was observed when the TS% increased from 0.5% to 2%. The slope of the curve then decreased drastically and the required specific energy to achieve a certain DD was almost constant regardless of the increase in TS%.

Using $COD_{solubilisation}$ and plotting the results with respect to DD (Fig. 3), a perfect linear relationship with an $R^2 = 1.0$ was obtained. Thus, $COD_{solubilisation}$ from now on can be used to evaluate the solubilisation degree as it proved to be an accurate and easy measure. In Table 1, the maximum DD is 42.6% at a specific energy of 30,000 kJ kg^{-1} TS^{-1} compared to Tiehm et al., (2001) and Rai et al., (2004) who applied almost double this specific energy on WAS to achieve a DD of 40% and 25%, respectively. The maximum solubilisation of hog manure measured as $COD_{solubilisation}$ was 27.3%, whereas Bougrier et al., (2005) and Khanal et al., (2006), using WAS, achieved 41.6% and 16.2% of SCOD/ TCOD at specific energies of 14,547 and 66,800 kJ kg^{-1} TS^{-1}, respectively. Applying ultrasonication on hog manure at different specific energy inputs achieved an increase of 1.35 mg SCOD kJ^{-1} kg^{-1} TS^{-1} compared to 0.12, 0.15, 0.45 and 0.9 mg SCOD kJ^{-1} kg^{-1} TS^{-1} calculated from the data reported by Gronroos et al., (2005); Khanal et al., (2006); Navaneethan, (2007); Bunrith, (2008), respectively indicating greater pre-treatment potential of hog manure by ultrasonication compared to WAS.

3.5. Methane production

The biochemical methane potential (BMP) test was used to evaluate anaerobic biodegradability in batch reactors. Fig. 4 shows the cumulative methane production with the digestion period at different sonication energy inputs with data summarized in Table 2. As shown in Fig. 4, no lag phase was observed due to sufficient soluble substrates. With respect to the results in Table 2, it is clearly observed that ultrasonication of hog manure enhanced the biogas production at low energy inputs compared to unsonicated hog manure. Methane potential increased by 28% for a specific energy input of 500 kJ kg^{-1} TS^{-1}, while the increase at high energy inputs (30,000 kJ kg^{-1} TS^{-1}) was only 20.7%. While the % increase in methane production rate increased by increasing the energy input, maximum increase in methane production rate was 80.6% compared to unsonicated hog manure at specific energy input of 30,000 kJ kg^{-1} TS^{-1}. The increase in methane production rate for specific energy input of 500 kJ kg^{-1} TS^{-1} (high methane potential) was about 61.3%. Therefore, specific energy input of 500 kJ kg^{-1} TS^{-1} can be considered to be the suitable energy input for anaerobic digestion of ultrasonicated hog manure. On the other hand, the reported optimum specific energy for ultrasonic pre-treatment of WAS in the literature was significantly higher than what has been achieved in this work, e.g., an optimum energy of 11,000 and 12,000 kJ kg^{-1} TS^{-1} were reported by Wang et al., (2006) and Navaneethan, (2007), respectively. The final concentration of VSS in the ultrasonicated hog manure reactor was 14% lower than that of the unsonicated hog manure reactor.

An economic analysis was conducted and the results are summarized in Table 2, which shows that the ultrasonication energy input per ton of TS, as well as the value of the methane produced based on power and natural gas costs of \$0.07 $(kWh)^{-1}$ and \$0.28 m^{-3}, respectively. Thus, the cost analysis also indicates that a specific energy input of 500 kJ kg^{-1} TS^{-1} can be considered to be the optimum energy for the pretreatment of hog manure by ultrasonication to be economically viable.

4. Conclusions

Based on the finding of this study, the following conclusions can be drawn:

- For hog manure, the disintegration of particles by ultrasonication was more pronounced for the smaller sizes, i.e., in the 0.6 to 60 μm range.
- Reduction of VS by ultrasonication increased with increasing power input in the 500-5,000 kJ kg^{-1} TS^{-1} and reached a plateau at ≥10,000 kJ kg^{-1} TS^{-1}.
- The $COD_{solubilisation}$ correlated very well with the DD.

- At solids content of 2%, the specific energy input increased from 10,000 to about 30,000 kJ kg^{-1} TS^{-1} for an additional 15% increase in degree of disintegration, whereas at TS of about 9%, the specific energy input increased from 250 to about 3,300 kJ kg^{-1} TS^{-1} to achieve the same increase in DD. Therefore, ultrasonication is more effective pretreatment process for sludge with higher TS content.
- The optimum specific energy input for methane production was 500 kJ kg^{-1} TS^{-1}, and resulted a 28% increase in methane production, subsequently about \$ 4.1 ton^{-1} of dry solids excess energy output.

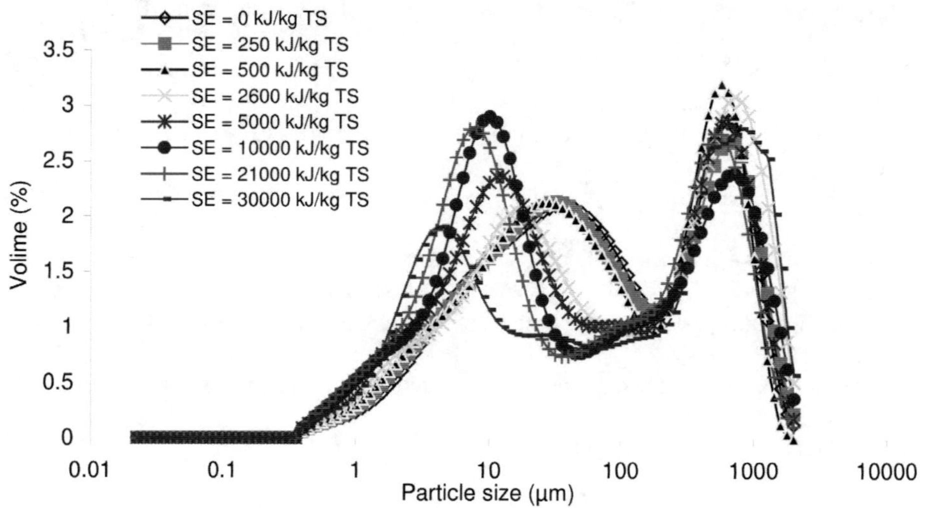

Fig. 1. Particle size distribution for different specific energy inputs.

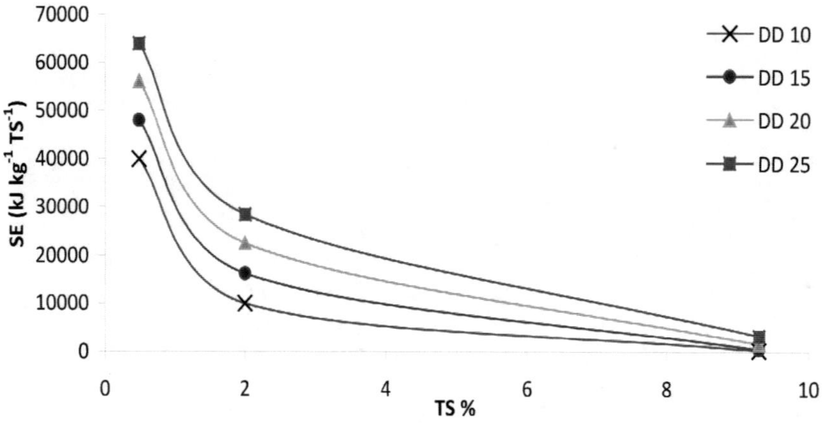

Fig 2. Specific energy input for different TS at different degree of disintegrations.
*Data in this graph are from this study, Tiehm et al. (2001), and Rai et al. 2004.

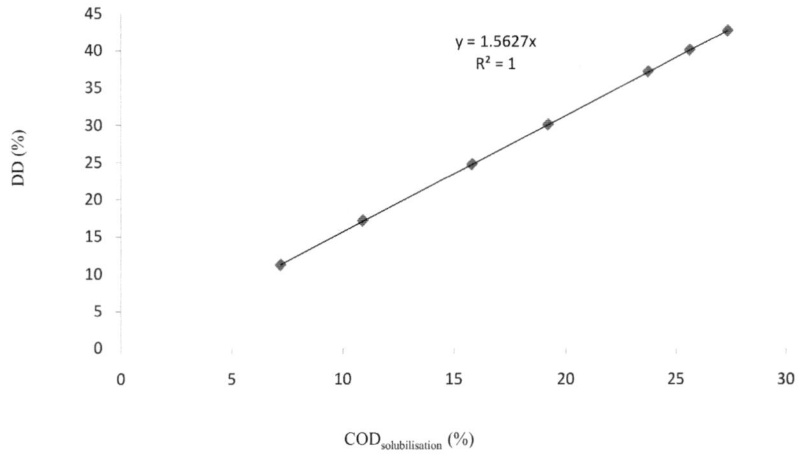

Fig. 3. Relationship between $COD_{solubilisation}$ and DD_{SCOD} (%)

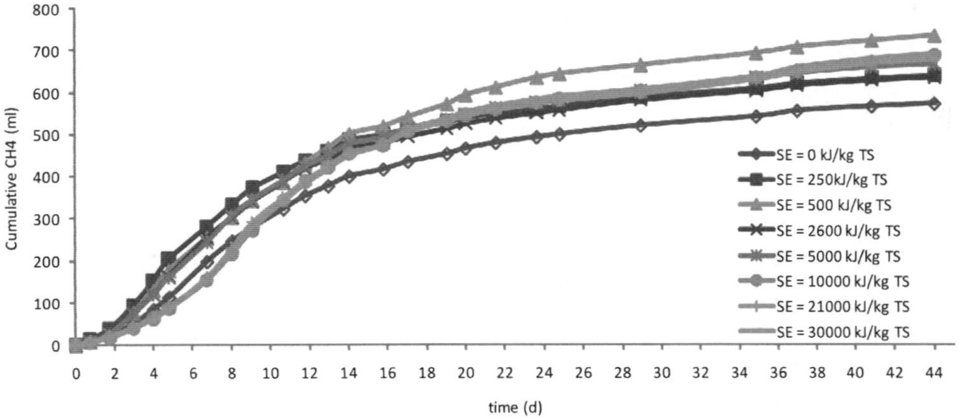

Fig. 4. Cumulative methane production for different specific energy inputs.

Table 1. Particle size information and COD solubilization for different specific energy inputs

SE (kJ kg^{-1} TS^{-1})	0	250	500	2500	5000	10000	21000	30000
d_{50} (µm)	59.0	56.0	53.9	47.3	39.7	33.3	27.4	12.2
SSA ((µm^2 g^{-1})	0.52	0.56	0.59	0.63	0.78	0.8	0.91	1.2
% Reduction in VS	-	5.2	19.9	23.7	23.8	29.6	30.4	32.0
DD (%)	-	11.2	17.1	24.7	30	37.1	40	42.6
$COD_{solubilisation}$ (%)	-	7.2	10.9	15.8	19.2	23.7	25.6	27.3

d_{50}: 50% of particles volume having a diameter lower than or equal to d_{50}.

Table 2. Sonication and Methane Energy per ton of TS

	Methane		Power input		Methane out		
SE (kJ kg^{-1} TS^{-1})	% Increase in methane potential	% Increase in methane production rate	kWh/ton TS_{in}	Price $/ ton TS_{in}	Increase of CH_4 (mL)	CH_4 m^3/ ton TS_{in}	Price $/ ton TS_{in}
0	-	-	0	-	-	-	-
250	11.7	33.7	69	4.9	67	17.2	4.8
500	28.0	61.3	139	9.7	160	50.4	14.1
2,600	10.9	43.5	722	50.6	62	201.2	5.6
5,000	16.3	35.5	1389	97.2	93	29.3	8.2
10,000	19.9	46.6	2778	194.4	114	37.9	10.6
21,000	18.7	75.4	5833	408.3	107	36.3	10.2
30,000	20.7	80.6	8333	583.3	118	40.0	11.2

References

APHA, 1998. Standard Methods for the Examination of Water and Wastewater, 20th ed.: American Public Health Association, Washington DC, USA.

Bougrier, C., Carrère, H., Delgenès, J.P., 2005. Solubilisation of waste-activated sludge by ultrasonic treatment: Chem. Eng. J. 106, 163-169.

Bunrith, S., 2008. Anaerobic digestability of ultrasound and chemically pretreated waste activated sludge: Master thesis study, Asian Institute of Technology, Bangkok, Thailand.

Gonze, E., Pillot, S., Valette, E., Gonthier, Y., Bernis, A., 2003. Ultrasonic treatment of an aerobic activated sludge in a batch reactor: Chem. Eng. Process. 42, 965–975.

Gronroos, A., Kyllonen, H., Korpijarvi, K., Pirkonen, P., Paavola, T., Jokela, J., Rintala, J., 2005. Ultrasound assisted method to increase soluble chemical oxygen demand (SCOD) of sewage sludge for digestion: Ultrason. Sonochem. 12, 115–120.

Higgins, M., Glindemann, D., Novak, J.T., Murthy, S.N., Gerwin, S., Forbes, R., 2004. Standardized biosolids incubation, headspace odor measurement and odor production consumption cycles: proceedings Water Env. Federation and AWWA Odors and Air Emissions Conference, Bellevue, Washington.

Khanal, S.K., Isik, H., Sung, S., Van Leeuwen, J., 2006. Ultrasound pretreatment of waste activated sludge: evaluation of sludge disintegration and aerobic digestion: In CD-ROM Proceedings of IWA World Water Congress and Exhibition, Sep 10-14, Beijing, China.

Lowry, O.H., Rosebrough, N.H., Farr, A.L., Randall, R.J., 1951. Protein measurements with the folin phenol reagent: J. Biol. Chem. 193, 265–275.

Muller, J., Pelletier, L., 1998. Désintégration mécanique des boues activées, L'eau, l'industrie, les nuisances. 217, 61–66. (as reported in Schmitz, et al., 2000)

Navaneethan, N., 2007. Anaerobic digestion of waste activated sludge with ultrasonic pretreatment: Master thesis study, Asian Institute of Technology, Bangkok, Thailand.

Pavlostathis, S.G., Gosset, J.M., 1986. A kinetic model for anaerobic digestion of biological sludge: Biotechnol. Bioeng. 27, 1519–30.

Pérez-Elvira, S.I., Nieto Diez, P., Fdz-Polanco, F., 2006. Sludge minimisation technologies: Reviews in Environmental Science and Biotechnology, 5, 375–398.

Rai, C.L., Struenkmann, G., Mueller, J., Rao, P.G., 2004. Influence of ultrasonic disintegration on sludge growth reduction and its estimation by Respirometry: Environ. Sci. Technol. 38, 5779-5785.

Schmitz, U., Berger, C.R., Orth, H., 2000. Protein analysis as a simple method for the quantitative assessment of sewage sludge disintegration: Water Res. 34(14), 3682–3685.

Tiehm, A., Nickel, K., Zellhorn, M., Neis, U., 2001. Ultrasonic waste activated biosolid disintegration for improving anaerobic stabilization: Water Res. 35, 2003–2009.

Advancement of Membrane Technologies for Water Sustainability:Membrane Distillation and Forward Osmosis (invited)

T.S. Chung, Chemical and Biomolecular Engineering, National University of Singapore, Singapore, Singapore

Water, energy, affordable healthcare, and global warming are four major concerns globally resulting from resource depletion, record high oil prices, clean water shortages, high costs of pharmaceuticals, increased aging population, and changing climate conditions. Among many potential solutions, advance in membrane technology is one of the most direct, effective and feasible approaches to solve these sophisticated issues. Membrane technology is a fully integrated science and engineering which consists of materials science and engineering, chemistry and chemical engineering, biomedical and life science, separation and purification phenomena, statistical mechanics-based molecular simulation, process and product design. In this presentation, we will (1) review and summarize the recent progresses of emerging technologies on desalination and water reuse such as membrane distillation and forward osmosis, and (2) the molecular engineering of membrane materials and fabrication to produce high performance membranes.

Email: chencts@nus.edu.sg Websites: http://www.chee.nus.edu.sg/people/faculty_chungneal.html
http://www.chbe.nus.edu.sg/membraneresearch/

See more of this Session: International Forum On Sustainable Energy & Environment
See more of this Group/Topical: Topical G: Innovations of Green Process Engineering for Sustainable Energy and Environment

Perspective of Green Process Engineering in China (invited)

Suojiang Zhang, State Key Lab of Multi-phase Complex Systems, Institute of Process Engineering, Chinese Academy of Sciences, Beijing, China

Energy and environmental sustainability is vital for China's economic growth. In this presentation, the author will discuss (i) current challenges of chemical process industries in China, (ii) progress of green process engineering in transforming green chemistry/science into practical impacts, and (iii) innovation opportunity in areas of traditional process, new energy and high value products.

See more of this Session: International Forum On Sustainable Energy & Environment
See more of this Group/Topical: Topical G: Innovations of Green Process Engineering for Sustainable Energy and Environment

Collaborative Initiatives of Pacific Northwest National Lab with Chinese Academy of Science On Biomass Conversion and Utilization (invited)

Yong Wang, Pacific Northwest National Laboratory, Richland, WA

USA and China are major energy consumers and biomass producers. Production of chemicals, fuels, and energies from renewable biomass in these two countries has significant global impact to reduction of greenhouse gas emission and oil import dependence. In this technical session, I will talk about some experiences and success of collaborative research activities in biomass conversion area PNNL has worked with Chinese Academy of Science for recent years.

See more of this Session: International Forum On Sustainable Energy & Environment
See more of this Group/Topical: Topical G: Innovations of Green Process Engineering for Sustainable Energy and Environment

Economic and Policy Drivers of Research and Development in Green Process Engineering in China and the United States

Kira JM Matus, Sustainability Science Program, Harvard Center for International Development, Harvard University, Cambridge, MA

There is an increasing demand for greener chemical processes in both China and the United States. However, the difference in the economic and policy contexts of these two nations results in appreciable differences in the focus and progress of research and development efforts. Extensive field research of green chemistry and engineering in China and the United States has led to six key findings in three main categories. These findings are:

I. Globalization

1. Green chemistry and engineering innovations face additional barriers (compared to other innovations) because their costs and benefits are particularly "leaky" across spatial and temporal boundaries.

2. The system through which these innovations are developed are highly transnational.

3. Regardless of local contexts, technical barriers present a common challenge to the development and implementation of green engineering technologies.

II. Context

4. The different human resources and human capabilities found in any particular country context, as well as economic organization (how firms are structured, what is being produced) impacts whether, how, and in what ways these innovations occur.

III. Intervention Strategies

5. Efforts to green chemistry and engineering innovations need to include analysis of the where in the local and global innovation system interventions are most effective.

6. Successful green chemistry and engineering innovation requires smart, strategic engagement between different combinations of stakeholders.

These findings are illustrated by a comparison of the main economic and policy drivers of green chemistry and engineering innovations, the key barriers to the development and implementation of these innovations, and the different models of collaboration that are emerging in response. Finally, they lead to a set of key policy implications at the national and international level.

See more of this Session: International Forum On Sustainable Energy & Environment
See more of this Group/Topical: Topical G: Innovations of Green Process Engineering for Sustainable Energy and Environment

Key Consideration to Meet Increasing Demand of Green Process in China

Xin Xiao, Institute of Process Engineering, Beijing, China

Kira Matus, Harvard University, John F. Kennedy School of Government, Mclean,, VA

The pollution from rapid economy development in past 30 years in China has caused increasing demand on implementation of green process. To support Chinese sustainable development, the key consideration to meet this demand is how to promote Green Process research development in China and simultaneously accelerate its commercialization process. There appears to be more challenges arising from overcoming the engineering barriers in commercializing the research results from lab scale to large scale. Through some cases of rapid development of unconventional green mediums realizing clean productions, we try to conclude successful elements which would be useful for future development.

See more of this Session: International Forum On Sustainable Energy & Environment
See more of this Group/Topical: Topical G: Innovations of Green Process Engineering for Sustainable Energy and Environment

A Study of Municipal Solid Waste in US and Its Utilization as Transportation Fuel or Energy Resource

Wei Liu, Energy & Environmental, Pacific Northwest National Lab, Richalnd, WA

Mark Bearden, Energy & Environmental, Pacific Northwest National Lab, Richalnd, WA

Don Stevens, Energy & Environmental, Pacific Northwest National Lab, Richalnd, WA

United States produces the largest amount of municipal solid waste (MSW) in the world and its generation has been steadily growing. The amount of energy is equivalent to approximate 8~12 % of the energy consumed for transportation in US. Less than 10% of MSW is recovered for energy value in US, mostly by incineration. However, the incineration plant has low power production efficiency and is of environmental concerns. Gasification is viewed as an attractive process route for conversion of MSW into electricity and/or clean fuels. This presentation provides a critical review of MSW gasification technology development, current technology gaps, and research needs to enable more widespread use of MSW as a clean energy source or feedstock for liquid fuels.

See more of this Session: International Forum On Sustainable Energy & Environment
See more of this Group/Topical: Topical G: Innovations of Green Process Engineering for Sustainable Energy and Environment

Applications of Ionic Liquids in Green Process Engineering

Suojiang Zhang, Institute of Process Engineering, Chinese Academy of Sciences, Beijing, 100080, P. R. China, Beijing, China

Xiaoqian Yao, Institute of Process Engineering, Chinese Academy of Sciences, Beijing, 100080, P. R. China, Beijing, China

Xiaomin Liu, Institute of Process Engineering, Chinese Academy of Sciences, Beijing, 100080, P. R. China, Beijing, China

A significant amount of researches on ionic liquids (ILs) have been prompted from fundamentals to applications due to their unique characteristics, such as nonvolatility, excellent catalytic and dissolvability properties. For process engineering, the applications of ILs may improve reaction rate and selectivity, and reduce the cost of product separation, and therefore lead to great innovations in process technologies. We would like to use this opportunity to bring to the forefront the many exciting challenges and opportunities in this field.

However, there are still a number of challenges to be tackled for application of ILs in large scale. In order to design task-specific ILs and develop innovative green processes, it is indispensable to understand the nature of ILs from the multiscale viewpoints. Quantum mechanics (QM), molecular dynamics (MD) and computational fluid dynamics (CFD) have been performed to depict the multiscale structures and properties of ILs.

Based on the fundamental understanding in the structures, properties and process scale-up of ILs, several innovative green processes using ILs as catalysts or solvents, such as cleaner process of Methyl methacrylate (MMA), cleaner process of ethylene glycol and de-acid process from oils, were developed.

Here we firstly show green process for producing methyl methacrylate (MMA), in which ionic liquids as the solvents or catalysts lead to great innovations in the absorption and reaction processes. A reaction-separation coupled one-step technique using ILs as medium is developed for MAL to MMA without methacrylic acid (MAA) step of the existing processes. Comparing the traditional hydrocyanic acid process, the new process achieved both the economic and environmental benefits. Another example is using ILs as catalyst for producing ethylene glycol instead of the traditional direct hydration of ethylene oxide, the new process could save large quantity of energy and significantly reduce wastewater discharge amount. A new naphthenic acid extraction process based on the principle of "switchable ionic liquid" is also developed. Because of the difference of the polarity between ionic liquid and lube base oil, naphthenic acid could be removed and recovered from lube base oil. This technique will be a new high atomic economy process, no pollution and "zero" emission.

Acknowledgments

The authors sincerely appreciate financial supports from National Scientific Funding of China for Distinguished Young Scholar (20625618 and 20806083), National Basic Research Program of China (973 Program, 2009CB219902).

See more of this Session: Novel Catalytic and Separation Process Based On Ionic Liquids I
See more of this Group/Topical: Topical G: Innovations of Green Process Engineering for Sustainable Energy and Environment

Direct Catalytic Conversion of Cellulose to 5-Hydroxymethylfurfural (HMF)

Z. Conrad Zhang, KiOR, Pasadena, TX

Haibo Zhao, Huntsman Chemicals, Woodlands, TX

Heather Brown, Pacific Northwest National Lab, Richland, WA

Yu Su, Pacific Northwest National Lab, Richland, WA

Guosheng Li, Pacific Northwest National Lab, Richland, WA

Xiao-Dong Zhou, Pacific Northwest National Lab, Richland, WA

Xiwen Huang, BASF, Ledgewood, NJ

Johnathan E. Holladay, Pacific Northwest National Lab, Richland, WA

Jim Amonette, Pacific Northwest National Lab, Richland, WA

John Fulton, Pacific Northwest National Lab, Richland, WA

Glucose and fructose, two abundant C6 sugar molecules from the nature, are potential feedstocks to provide sustainable supplies of basic chemical building blocks. 5-Hydroxymethylfurfural (HMF) is an important versatile sugar derivative and is a key intermediate between bio-based carbohydrate chemistry and petroleum based industrial organic chemistry. High sugar conversion to hydroxymethylfurfural (HMF) was achieved in ionic liquids with or without the addition of Bronsted acids. Catalytic amounts of metal halides in ionic liquids significantly reduced the fructose dehydration barrier, producing HMF in high yields. Most remarkably, glucose, a common sugar molecule, was selectively converted to HMF in good yield. Only negligible amount of levulinic acid was formed as a by-product. A possible mechanism is proposed for the unique catalytic function of $CrCl_2$ in glucose conversion. Direct conversion of cellulose to HMF in a single process represents an ultimate challenge because it involves multiple elementary steps such as decrystallization and depolymerization of cellulose in high yield to glucose and the consecutive conversion of glucose to HMF. A unique catalytic system was successfully designed that led to a breakthrough toward this goal. Preliminary results will be presented and discussed.

See more of this Session: Novel Catalytic and Separation Process Based On Ionic Liquids I
See more of this Group/Topical: Topical G: Innovations of Green Process Engineering for Sustainable Energy and Environment

An Ionic Liquid-Based Process for Conversion of Biomass Into 5-Hydroxymethylfurfural (HMF) as Petrochemical Intermediate

Wei Liu, Energy & Environmental, Pacific Northwest National Lab, Richalnd, WA

Allan Cooper, Pacific Northwest National Lab, Richland, WA

Richard Zheng, Pacific Northwest National Lab, Richland, WA

Heather Brown, Pacific Northwest National Lab, Richland, WA

Most of today's bulk petrochemical and chemical products are manufactured from raw materials that are derived from petroleum oil or natural gas. For example, aromatics are produced from crude oil through a series of energy and capital-intensive refining process steps. Replacement of such oil-based feedstock with intermediates derived from renewable biomass material has both energy and environmental benefits by reducing oil consumption and green house gas emissions. In addition, chemical feedstock typically has a higher value than the liquid transportation fuel. 5-hydroxymethylfurfural (HMF) has been proposed as a flexible platform chemical to substitute for today's petroleum-derived feedstock such as paraxyelene in manufacturing of polyester and polyamide. A good amount of research has been reported in the literature for synthesis of HMF from renewable hydrocarbon sources. However, there are no practical processes yet to produce HMF at a cost level competitive with today's petrochemical feedstock. A continuous flow process based on ionic liquids will be described in this presentation for conversion of sugars, cellulose into HMF. The ionic liquid fluid serves both as a green solvent to dissolve the sugar-type biomass and as a catalyst carrier. Fundamental studies on reaction kinetics and HMF/ionic liquid separation will be discussed pertinent to reactor design and ionic liquid recovery for development of a low-cost HMF manufacturing process.

See more of this Session: Novel Catalytic and Separation Process Based On Ionic Liquids I
See more of this Group/Topical: Topical G: Innovations of Green Process Engineering for Sustainable Energy and Environment

Homogeneous Catalytic Conversion of Cellulose in to 5-Hydroxymethylfurfural (HMF) Using 1-Ally-3-Methylimidazolium Chloride (AmimCl) Ionic Liquids

Jian Zhang, ,Key Laboratory for Green Chemical Technology of Ministry of Education, School of Chemical Engineering & Technology, Tian Jin University, Beijing, China

Yan Cao, Key Laboratory of Green Process and Engineering, Institute of Process Engineering,Chinese academy of science, Beijing, China

Huiquan Li, Key Laboratory of Green Process and Engineering, Institute of Process Engineering,Chinese academy of sciences, Beijing, China

Xinbin Ma, ,Key Laboratory for Green Chemical Technology of Ministry of Education, School of Chemical Engineering & Technology, Tian Jin University, Tian jin, China

Abstract

The consumption of petroleum has surged during the 20th century because of the rise of the automobile industry. Unfortunately, the growing demand for fossil fuel resources comes at a time of diminishing reserves of these non-renewable resources, such that the worldwide reserves of oil are sufficient to supply energy and chemicals for only about another 40 year, causing widening concerns about "energy crisis". Biomass can serve as a source for both energy and carbon, and being renewable it is the only sustainable source of energy and organic carbon for our industrial society.

Moreover, production of energy from biomass has the potential to generate lower greenhouse gas emissions compared to the combustion of fossil fuels, because the CO2 released during energy conversion is consumed during sub-sequent biomass regrowth..

Biomass includes various plant components, such as starch and cellulose, that are originally formed by photosynthesis. As the most abundant source of biomass , cellulose is currently regarded as a promising alternative for fossil fuels as it cannot be digested by human beings and thus its use, unlike corn and starch, will not impose a negative impact on food supplies. During the last few decades, much effort has been devoted to increasing the utilization of cellulose. However, the utilization of cellulose has been limited to lumber, fuel, textile, paper, plastic, and so on because of it is resistant to degradation. Recently, a great deal of effort has been put toward the degradation of cellulose with enzymes, mineral acids, base, and superitical water.

One of the most attractive utilization of cellulose is its direct conversion into 5-hydroxymethylfurfural (HMF), which is a versatile and key intermediate in biofuel chemistry and petroleum industry. HMF and its derivatives can replace key petroleum-based building block.

According the literature, HMF is now prepared mainly from fructose using acid as catalyst . However, fructose is so expensive that arising the production cost of HMF, thus limits the availability and use of HMF industrially. Researchers are applying their energies to converting glucose to HMF, and have acquired satisfying results. Using a system containing chromium chloride and ionic liquids, the HMF yield is up to a record high of 68%. Nevertheless, this technology is still based on food resources such as starch.. thus, it is meaningful to exploring a method that can directly convert cellulose to HMF.

Because of the high crystallinity of cellulose, the conversion of cellulose into valuable chemicals is inefficient and uneconomical. The conversion condition is always high pressure and high temperature, yet the hydrolysis is slow and complicated, producing additional by-products and increasing the cost of product purification.

Studies by Rogers et al showed that the ionic liquid (IL) 1-butyl-3-methylimidazomium chloride ([C4mim]Cl) is a powerful solvent for cellulose. Since then solubilization of cellulose as well as the reaction of cellulose in ILs has attracted much attention. A lot of researches about the reaction of cellulose in ILs have been doing all over the world, including derivatization, hydrolysis, hydrogenation and hydrogenolysis.

The hydrolysis of cellulose in ILs have been done systemly using mineral acids, solid acids as catalysts. But neither of these researches are seting HMF as a target product. An USA patient described a method to prepare HMF from cellulose in the [C4mim]Cl ILs using chromium chloride as catalyst, but it's lack of few details.

The IL, 1-ally-3-methylimidazolium chloride ([Amim]Cl) has been proved to be a good solvent in dissolving cellulose. And it has been used in the acylation of cellulose from both the high quality pulp and agricultural residue.

Recently we are carrying out an experiment on hydrolysis of cellulose into HMF using [Amim]Cl as reaction medium. Primarily experimental results shows that with chromium chloride as catalyst, the HMF yield is up to 50%. And increasing the reaction time will result the decomposition of HMF. Further studies such as the mechanism of the hydrolysis, the kinetics of the hydrolysis are underway.

See more of this Session: Novel Catalytic and Separation Process Based On Ionic Liquids I
See more of this Group/Topical: Topical G: Innovations of Green Process Engineering for Sustainable Energy and Environment

Celluose acetate burytare prepared from sugarcane bagasse in ionic liquid

Kelin Huang[1,3], Yan Cao[2], Ben Wang[2], Huiquan Li[2*]

[1] Beijing University of Technology, Beijing 100124, China

[2] Key Laboratory of Green Process and Engineering, Institute of Process Engineering ,

Chinese Academy of Sciences, Beijing 100190, China

[3] Guangxi Research Institute of Chemical Industry, Nanning 530001, China

Abstract

Energy crisis is a huge challenge that mankind will face up to today, The statistic made by the International Energy Agency (IEA) on oil resources based on the production rate and oil reserves estimated that the main fossil fuels——petroleum, gas and coal for humanbeings can be exploited for only 40, 50 and 240 years respectively. Since petroleum resources in China are relatively poor, domestic petroleum production can not meet with the demands. Pertinent institution predicts that in 2020, 60% of China's petroleum demand will depend on import, and petroleum deficiency has become an unavoidable worldwide problem. So there is a need to increase the effciency of the plentifull and regenerated biomass utilization in order to promote the continuable development of China.

The agricultural residues represent an abundant, inexpensive, and readily available source of renewable lignocellulosic biomass for the production of environmentally friendly industrial products and have received increasing attention. It usually contains more than 30% cellulose, and is of limited commercial value at present. Currently China produces around 700 million tons of agricultural straw annually, however, most of it is simply burnt in the fields, creating a significant local air pollution problem. Therefore, effective utilization of these commonly available by-products as a source for cellulose will offer significant benefits to the economy and the environment. In addition , the cost of harvesting, handling, storing and

* Corresponding author. Tel.: +86-010-62621355; fax: +86-010- 62621355. *E-mail address:* hqli@home.ipe.ac.cn

transporting biomass resources is relatively low in most developing countries than developed countries. During the last few decades, much effort has been devoted to increasing the utilization of the agriculture waste for several end uses.

Among the various agricultural crop residues, sugarcane bagasse (SB) is the most abundant agricultural material in Southern China. SB is a fibrous residue of cane stalks left over after the crushing and extraction of the juice from the sugarcane. Nowadays, most of the SB is burnt to generate energy for the mills, but even then there is a considerable amount of it still wasted. For this reason, several papers have been produced setting alternatives for recycling it. Since SB consists of approximately 50% cellulose, which makes it possible to use this residue by producing cellulose derivatives.

As we know that cellulose is the most abundant of all natrally occurring substances and widely used as a raw material in numerous industrial applications. However Due to its high crystallinity, cellulose is insoluble in most of the traditional organic solvents, what limitate significantly its applicability at its original form. An alternative path for increasing the cellulose applicability is chemically modifying it by incorporating many kinds of chemical groups on the hydroxyl groups of the glycosidic units. According to this procedure, up to now several cellulose derivatives have already sucessfully produced from sugarcane bagasse cellulose (SBC), as for example, cellulose phthalates, cellulose succinates, cellulose sulphate and so on.

Among cellulose derivatives, cellulose acetate butyrate (CAB) is one of the most important due to their multiple applications in the field of modern coatings, membranes, plastic, osmotic drug delivery and so on. The molecular structure of CAB is shown in Figure1. The properties and practical applications of CAB depend on the butyryl content, which determines their compatibility with other plastics and resins, as well as their solubility in many solvents. CAB with relatively high butytyl content exhibited good solubility in organic solvents, while being more hydrophobic than CA. Therefore, it is recognized that the appearance of the mixed-cellulose ester provide the feasibility for improving the properties of cellulose esters.

Normally, CAB with full degree of substitute (DS) are prepared in heterogeneous medium (Figure 2), using excess of acetic anhydride and butyric anhydride in the presence of sulfuric acid as catalysis. CAB with a specific DS value lower than three can be obtained by hydrolysis. However, this method always result in the serious degradation of cellulose, uncontrollable structure of CAB and environmental pollution.

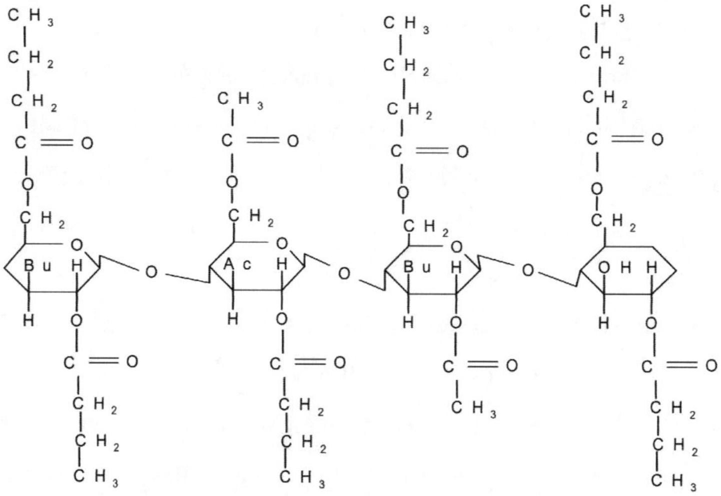

Figure 1. Molecular structure of cellulose acetate butyrate

A major stimulation for the investigation of homogeneous paths for cellulose functionalization was the discovery of a variety of new solvents for cellulose with its extended supra-molecular structure. The interest in the derivatization of cellulose under homogeneous conditions resides mainly in the fact that it allows a more efficient control of the DS of the cellulose derivative through adjustment of the molar ratio of the derivatizing agent and cellulose. Moreover, the substituting groups can be introduced more regularly along the polymer chain and materials with well-defined characteristics can be produced using methods with good reproducibility

Up to now, several direct solvents are found suitable for chemical functionalization reactions of the cellulose. These solvents include N, N-dimethylacetamide(DMAc)/LiCl, dimethyl sulfoxide(DMSO)/tetrabuthlammonium fluoride trihydrate (TBAF) and some molten salt hydrate, such as $LiClO_4 \cdot 3H_2O$, and $LiSCN \cdot 2H_2O$. However, the drawbacks in the case of above solvent systems are toxicity, high cost, difficulty in solvents recovery, severe side-reaction, and instability during cellulose derivatization. At the same time, the cellulose esters reported in the literature are mostly the single-cellulose esters, such as CA、CP、CB and so on. While the mix-ester was only mentioned in DMAc/LiCl solvent system with few detail synthesis nor characterizaiton data.

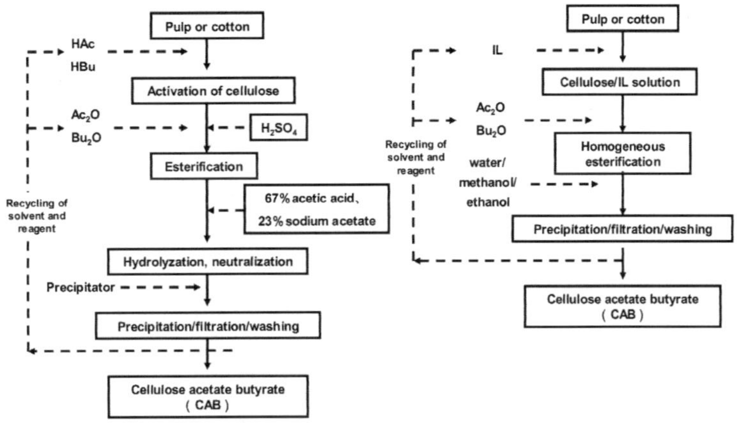

Figure2. Schematic flow for producing CAB (left: traditional heterogeneous process; right: homogeneous process in ionic liquid.

Recently, room-temperature ionic liquids (ILs), being considered as green solvents, have been successfully used to dissolve and process cellulose. As a novel non-derivatizing cellulose solvent, ILs have also been found to be the promising reaction media for cellulose derivatization. The IL, 1-allyl-3-methylimidazolium chloride (AmimCl) has been used as reaction media for acylation of cellulose from both the high quality pulp and agriculture residue.

Recently we are carrying out an experiment on synthesize CAB from SBC in AmimCl homogeneously. Preliminary experimental results are shown in Table 1, it shows that by a sequence of first adding butyric anhydride and then acetic anhydride in the cellulose/AmimCl solutions, the butyryl contents of the CAB prepare from SBC can reach as high as 46 wt %, and all the CAB sample in Table 1 can dissolve in acetone. While ^{1}H NMR spectra in Figure 3 used to confirm the structure of CAB, the identification of the hydrogen for the signal at δ = 1.0 ppm was attributed to the CH_3 from the butyryl; the signal at δ = 1.5 ppm and 2.2 ppm attributed to the CH_2 both from the butyryl; δ = 2.0 ppm attributed to the CH_3 from acetyl; δ = 3 ~ 5.5ppm attribute to the hydrogen from the cellulose backbone. Further researches such as the thermal stability, distribution of the substitution and viscosity are underway.

Table 1 Homogeneous synthesis of CAB in AmimCl (conditions and results)

No.	Cellulose	Reaction conditions[a]					Reaction results[b]				
		Temp (℃)	M1 (mol/mol)	t 1 (h)	M2 (mol/mol)	t 2 (h)	DS_A	A%	DS_B	B%	DS_{total}
CAB1	SBC	100	3	1	5	3	1.26	19	1.00	25	2.26
CAB2	SBC	100	5	1	5	3	1.25	18	1.23	29	2.48
CAB3	SBC	100	5	1	3	3	0.69	11	0.96	26	1.65
CAB4	SBC	100	5	3	5	3	0.63	9	1.66	39	2.29
CAB5	SBC	100	5	0.5	5	3	1.03	16	1.11	28	2.14
CAB6	SBC	100	9	1	3	3	0.73	11	1.39	34	2.12
CAB7	SBC	100	9	3	5	3	0.55	8	1.59	38	2.14
CAB8	SBC	100	9	5	5	1	1.17	17	1.34	31	2.50
CAB16	SBC	100	13	5	5	1	0.30	4	2.07	46	2.36
CAB_M	MC	100	5	3	5	3	0.27	4	2.15	47	2.42

[a] : M1 - Bu2O/AGU (mol/mol) ; M2 - Ac2O/AGU (mol/mol)

[b] : A% - acetyl content; B% - butyryl content

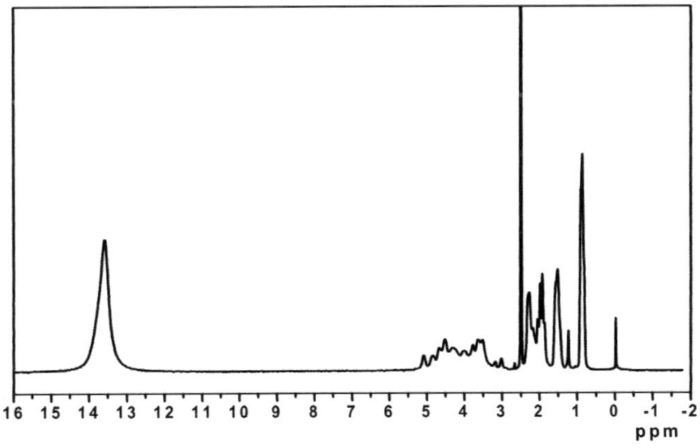

Figure3. ^{1}H NMR spectrum of CAB 16

Evaporative Separation of Ionic Liquid Following Biomass Pretreatment: Generation of Required Thermodynamic Data

Thehazhnan K. Ponnaiyan, Chemical & Environmental Engineering, University of Toledo, Toledo, OH

Anantharam P. Dadi, SuGanit Systems, Toledo, OH

Constance A. Schall, Chemical & Environmental Engineering, University of Toledo, Toledo, OH

Sasidhar Varanasi, Chemical & Environmental Engineering, University of Toledo, Toledo, OH

Plant (lignocellulosic) biomass is an abundant, inexpensive and sustainable source of organic carbon, for the production of renewable fuels and chemicals. However, many technical and environmental challenges exist in developing commercially viable processes for producing sugars from biomass that serve as precursors for fuel and chemical production. Pretreatment and enzymatic hydrolysis are critical steps, and most of the current pretreatment methods do not effectively disrupt the biomass structure, especially the crystallinity of cellulose, which is a major barrier for efficient enzymatic hydrolysis. Ionic liquids (ILs) being non volatile, non-flammable, recyclable and designer friendly, are gaining wide recognition as green solvents, and their unique solvating properties make them ideal for pretreating lignocellulosic substrates. Enhanced enzymatic hydrolysis yields of glucan to glucose and xylan to xylose are observed for ionic liquid pretreated ligocellulosic substrates like corn stover and poplar. In IL-pretreatment, the IL is displaced from the biomass by washing the pretreated biomass with an anti-solvent (water/ethanol). The IL preferentially dissolves in the anti-solvent and hence can be completely recovered. However, from an economic stand-point it is important that the IL is separated from the IL/water mixture. This is also desirable to avoid any ecological consequences associated with the presence of IL in wash streams and be able to reuse the process water. This work addresses: (1) the pretreatment of lignocellulosic biomass using suitable ionic liquids, and (2) a facile thermo gravimetric method for efficiently generating the required thermodynamic (vapor liquid equilibrium) data needed for designing evaporative separation of IL from the anti solvent.

See more of this Session: Novel Catalytic and Separation Process Based On Ionic Liquids I
See more of this Group/Topical: Topical G: Innovations of Green Process Engineering for Sustainable Energy and Environment

Mixed Matrix Membrane Based On Room Temperature Ionic Liquid for Gas Separation

Yeny C. Hudiono, Chemical and Biological Engineering, University of Colorado, Boulder, CO

Rich Noble, University of Colorado, Boulder, CO

Douglas Gin, Department of Chemical and Biological Engineering, University of Colorado, Boulder, CO

Polymerizable ionic liquids (poly (RTILs)) - ionic liquids (RTILs) composite membranes were previously fabricated, and their gas transport were investigated. The addition of 20 mol % of RTILs in poly (RTILs) membrane increases the CO_2, CH_4, and N_2 permeability of the composite membrane due to more rapid gas diffusion in the RTILs 1, 2. However, the presence of RTILs decreases the CO_2/CH_4 selectivity. On the contrary, previous studies have shown that mixed matrix membranes consisting of polymeric matrix and dispersed inorganic particles have the potential for outstanding separation properties, but some challenges such as a good adherence at the inorganic/organic interface still remain uncertain.

In our current study, we have successfully fabricated supported ionic liquid - zeolite membrane and showed an improvement in selectivity in comparison to neat supported ionic liquid membrane (SILM). The addition of only 10 wt. % of SAPO-34 particles increases the CO_2/CH_4 and CO_2/N_2 selectivity by 69% and 65% respectively. Moreover, we have also fabricated a three component membrane that is composed by poly (RTIL) - RTIL - and zeolite materials without the presence of additional solvent. We showed that addition of 10 wt. % SAPO-34 increases CO_2 permeability 63%, and CO_2/CH_4 and CO_2/N_2 selectivity by 11%. We hypothesize that the presence of RTILs can serve as a "wetting" media to better facilitate a good adhesion between zeolite and poly (RTIL) interface.

In this study, we extend our study to investigate the effect of anion of RTILs and their interaction with zeolite particles on gas separation performances. We will also analyze how the RTILs anion will affect the adhesion between poly (RTIL) and zeolite particle, and improve gas separation performances.

1. Bara, J. E.; Gin, D. L.; Noble, R. D., Effect of Anion on Gas Separation Performance of Polymer-Room-Temperature Ionic Liquid Composite Membranes. Ind Eng Chem Res 2008, 47, (24), 9919-9924.

2. Bara, J. E.; Hatakeyama, E. S.; Gin, D. L.; Noble, R. D., Improving CO2 permeability in polymerized room-temperature ionic liquid gas separation membranes through the formation of a solid composite with a room-temperature ionic liquid. Polym Advan Technol 2008, 19, (10), 1415-1420

See more of this Session: Unconventional Technologies for CO2 Capture, Conversion and Utilization I
See more of this Group/Topical: Topical G: Innovations of Green Process Engineering for Sustainable Energy and Environment

Liquid Membrane Immobilized in Durable Inorganic Matrix for CO2 Separation

Jian Zhang, Pacific Northwest National Lab, Richalnd, WA

Wei Liu, Energy & Environmental, Pacific Northwest National Lab, Richalnd, WA

In this work, we will report a new liquid/inorganic membrane product concept for CO2 separation. A liquid fluid of selective CO2 absorption properties is immobilized in a thin, inorganic substrate sheet (20~50um) of graded pore structures. CO2 selectively absorbs into the liquid fluid in the feed side, the absorbed species diffuse across the liquid film under a partial pressure gradient of CO2, and CO2 desorbs in the permeate side. The inorganic matrix provides mechanical strength and membrane surface area. Compared to the absorption process, CO2 continuously absorbs and desorbs in the same spot during membrane separation. Thus, energy and capital costs associated with regeneration in the conventional absorption process can be saved. Compared to the conventional membrane, the proposed membrane provides combined performance attributes of high membrane area packing density, low cost, and durability. CO2 permeance obtained with the proposed hybrid membrane is among the highest in the literature.

See more of this Session: Unconventional Technologies for CO2 Capture, Conversion and Utilization I
See more of this Group/Topical: Topical G: Innovations of Green Process Engineering for Sustainable Energy and Environment

Optimization of a CO_2 Capture System Using Chemical Absorption and Desorption

Swetha D, Department of Chemical Engineering, IIT Madras, Chennai, India

Sreenivas Jayanti, Department of Chemical Engineering, IIT Madras, Chennai, India

Capture and storage of carbondioxide (CO_2) from fossil fuel fired power plants is drawing increasing interest as a potential method for the control of greenhouse gas emissions. An optimization and technical parameter study for a CO_2 capture process of the flue gas of a commercial gas power plant, based on absorption/desorption process with different solvents, using Aspen Plus has been performed. The optimization has the aim of reducing the energy requirement for solvent regeneration (which is high in a CO_2 sequestration by absorption and later desorption) and also the reflux ratio needed. The simulations have been carried out using the ASPEN PLUS process simulation software for CO_2 compositions in the flue gas. ASPEN PLUS optimizes the absorber in terms of the minimum height of the absorber, minimum solvent loading, packing segments that is required to get more capture efficiency with proper packing properties that are in built within the software. The simulations considered the flue gas from a 600 MW coal-fired power plant with a flue gas mass flow rate of 577 kg/s with a CO_2 mass fraction of 15%. The simulations have been carried out for an MEA-based absorption/desorption system (Kohl, 1997) and for various compositions of piperazine-promoted potassium carbonate (K_2CO_3/PZ) mixtures (Cullinane, 2005) which are more environmentally friendly than mono ethanolamine (MEA)-based systems. The desorber pressure has also been varied to reduce the overall heat duty of the stripper (desorber) system. The principal features of the computed absorption system are summarized in Table 1. It can be seen that for the same capture rate of CO_2 coming from the same power plant, the K_2CO_3/PZ with the desorber operating at 3 bar gives a significant reduction (~25%) of the reboiler heat duty. The size of the absorbption/ desorption is also reduced with the K_2CO_3-based system.

Table 1. Comparison of the principal features of the CO_2 absorption system based on MEA and PZ-doped K_2CO_3

Feature	MEA based System	K_2CO_3/PZ system
Solvent concentration (wt%)	30	22.1/13.8
CO_2 capture rate (%)	90	90
Desorber pressure (bar)	2.1	3.0
Reboiler temperature (^oC)	128	125
Solvent flow rate (m^3 /tCO_2)	27.8	74.4
Reboiler heat duty (GJ/t CO_2)	3.3	2.4
Absorber Column height (m)	18	12
Absorber column diameter(m)	11	12.7

See more of this Session: Unconventional Technologies for CO2 Capture, Conversion and Utilization I
See more of this Group/Topical: Topical G: Innovations of Green Process Engineering for Sustainable Energy and Environment

Spectroscopic Observation of CO2 Absorption Behavior of Nanoparticle Ionic Materials

Youngjune Park, Earth and Environmental Engineering and Chemical Engineering, Columbia University, New York, NY

Kunyi Andrew Lin, Earth and Environmental Engineering and Chemical Engineering, Columbia University, New York, NY

Ah-Hyung Alissa Park, Earth and Environmental Engineering and Chemical Engineering, Columbia University, New York, NY

Ionic liquids, a molten salt composed of cations and anions, have received much attention due to their various physicochemical properties such as low melting point, low combustibility, non-volatility, excellent thermal stability, and solvating property, which allow them to be applied to various energy and environmental processes. Because ionic liquids could easily be modified by numerous combinations of cations coupled with anions, they possess tunable properties to meet the specific needs; for example, as potential green solvents and electrolytes. Recently, many researchers are investigating ionic liquids as novel CO2 capture medium. Compared to the conventional amine-based solvents (e.g., monoethnolamine (MEA)), which often suffer from corrosion and vaporization/degradation loss, ionic liquids exhibit reversible CO2 absorption and desorption without environmental problems because of its immeasurable vapor pressure. Therefore, ionic liquids have been suggested as an alternative option for CO2 capture technology.

In this study, we are focused on the newly synthesized ionic materials called Nanoparticle Ionic Materials (NIMS), which are nanoscale analogs of ionic liquids. NIMS consist of inorganic nano-sized cores and amine-functionalized corona. By tuning its functional groups as well as the volume fraction of nanoparticles, NIMS can be designed for the efficient CO2 capture with great flexibility for the counterion species. CO2 absorption and desorption behavior of NIMS are investigated through spectroscopic observation by employing IR and Raman spectroscopy. The results are compared with those of conventional and task-specific ionic liquids. The findings of this study would offer deeper understanding of CO2 absorption and desorption mechanisms of NIMS under various operating conditions.

See more of this Session: Unconventional Technologies for CO2 Capture, Conversion and Utilization I
See more of this Group/Topical: Topical G: Innovations of Green Process Engineering for Sustainable Energy and Environment

A New Paradigm of Anthropogenic CO2 Reduction: Adsorptive Fixation of CO2 From the Ambient Air as a Carbon Negative Technology

Sunho Choi, School of Chemical & Biomolecular Engineering, Georgia Institute of Technology, Atlanta, GA

Jeffrey H. Drese, School of Chemical & Biomolecular Engineering, Georgia Institute of Technology, Atlanta, GA

Peter Eisenberger, Global Thermostat, LLC

Christopher W. Jones, School of Chemical & Biomolecular Engineering, Georgia Institute of Technology, Atlanta, GA

In recent years, growing concern about the effect of anthropogenic CO2 on global climate change have motivated significant interest ain renewable energy technologies that are or can be carbon neutral, such as use of wind, solar and biomass-based energy. This is because the atmospheric CO2 concentration has risen about 35% to a value of 380 ppm since the industrial revolution due to widespread use of fossil fuels. Even in the face of significant growth of renewable energy usage, it is expected that the atmospheric CO2 concentration will continue to rise over the next decades. To this end, capture of CO2 from large point sources such as fossil fuel burning power plants has been proposed as a means to reduce anthropogenic CO2 emissions, with the captured CO2 sequestered permanently or semi-permenantly geologically, or used to some degree in the synthesis of fuels or chemicals. However, it should be noted that this approach can be at best carbon neutral, and as such, it does not offer the possibility to reduce atmospheric CO2 levels below where they are today. Furthermore, capture of CO2 from distributed sources such as cars, planes and ships appears economically unfeasible.

In contrast, CO2 capture from the ambient air can be carbon negative. This approach has been largely overlooked due to the perceived technical difficulty and cost associated with capturing CO2 from such a low concentration source. In this talk, a new concept of CO2 capture technique will be presented, which is capable of removing CO2 from the ambient atmosphere by means of the CO2 fixation on the solid adsorbent materials. Different types of solid sorbents have been prepared and tested evaluated for their adsorption capabilities using CO2 at a concentration of 400 ppm, to demonstrate the effectiveness of this idea. Experimental data obtained strongly suggest that this concept, i.e., CO2 capture from ambient air using solid sorbents, is a promising strategy that may be used to reduce the level of CO2 from the atmosphere.

See more of this Session: Unconventional Technologies for CO2 Capture, Conversion and Utilization I
See more of this Group/Topical: Topical G: Innovations of Green Process Engineering for Sustainable Energy and Environment

Highly Efficient Fischer-Tropsch Synthesis in Ionic Liquids

M. M. (Rien) Sinke, Process & Energy, Delft University of Technology, Delft, Netherlands

Maaike C. Kroon, Faculty of Mechanical, Maritime and Materials Engineering, Delft University of Technology, Delft, Netherlands

Cor J. Peters, Process and Energy, Delft University of Technology, Faculty of Mechanical, Maritime and Materials Engineering, Delft, Netherlands

Geert-Jan Witkamp, Section Process Equipment, Process & Energy Department, Faculty of Mechanical, Maritime and Materials Engineering, Delft University of Technology, Delft, Netherlands

Fischer-Tropsch synthesis involves production of liquid fuel from synthesis gas. Conventional Fischer-Tropsch processes are intrinsically inefficient as a result of the polymerisation nature of the reaction, which yields a product distribution rather than a single product. In the subsequent upgrading step the hydrocarbons that are too long are 'cracked' into molecules with shorter chain length, partly destroying what has been produced in the reaction step. This upgrading step is not only energy intensive, but it also accounts for a significant part of the capital expenditure.

In this work, a novel highly efficient Fischer-Tropsch process is developed that selectively produces hydrocarbons with the desired chain length (diesel) using an ionic liquid as solvent. This innovative process employs the phenomenon that the solubility of hydrocarbons in ionic liquids decreases with increasing hydrocarbon chain length. The polymerisation reaction takes place in a hydrophobic ionic liquid, which dissolves reactants, catalyst and hydrocarbons shorter than the desired product. However, at the moment the desired chain length is reached, the hydrocarbon is no longer soluble in the ionic liquid, and forms a separate layer that floats on top of the ionic liquid, thus avoiding contact with the catalyst and preventing further reaction. The product can be easily withdrawn from the reactor, abandoning the need for a conventional energy- and capital intensive upgrading section.

Process development results of the novel highly efficient Fischer-Tropsch synthesis in an ionic liquid will be presented at the conference.

See more of this Session: Novel Catalytic and Separation Process Based On Ionic Liquids II
See more of this Group/Topical: Topical G: Innovations of Green Process Engineering for Sustainable Energy and Environment

Regenerated Cellulose Films Prepared From Sugarcane Bagasse in Room Temperature Ionic Liquid

Yan Cao, Key Laboratory of Green Process and Engineering, Institute of Process Engineering,Chinese academy of science, Beijing, China

Ben Wang, Key laboratory of green process and engineering, Institute of Process engineering, Chinese Academy of Sciences, Beijing, China

Kelin Huang, The College of Materials Science and Engineering, Beijing University of Technology, Beijing, China

Huiquan Li, Key Laboratory of Green Process and Engineering, Institute of Process Engineering,Chinese academy of sciences, Beijing, China

Haitao Liu, Institute of Process Engineering,Chinese academy of sciences, Beijing, China

Abstract

Fossil based fuels, power, and products are critical to the current economy of the World. Continued massive consumption of fossil hydrocarbon reserves presents serious energy security and environmental quality issues for the World. Scientific and engineering breakthroughs currently made suggest the potential for transferring much of the current reliance on fossil hydrocarbons to bio-based fuels, power, and products that will do much to improve energy security, rural economic development, and environmental quality.

Cellulose is probably the most abundant biopolymer in the world. It is produced from regenerated resources and is a biodegradable polymer so that it will be one of the main chemical bio-based resources of the future. Its characterization is of particular interest to the textile, pulp, paper, and cellulose derivative industries. However due to the stiff molecule and close chain packing via numerous inter- and intra-molecular hydrogen bonds in cellulose, it is difficult to dissolve cellulose without chemical modification or derivatization. Up to now, Researchers have mainly used the viscose process which had lasted for 100 years and NMMO process which ranks among the new environmentally friendly processes to make regenerated cellulose. However, the viscose route brings about considerable pollution of air, watercourses, land, and current health fears over carbon disulphide air pollution. While, despite the industrialization of NMMO solvent system, it has some disadvantages associated with its use, such as the requirement for high temperature to dissolve, activation needed before dissolution, the degradation of cellulose, the side reaction of the solvent itself without an antioxidant, and its high cost as well. Therefore the search for excellent and environment friendly solvents for cellulose continues.

Recently, Room temperature ionic liquids have been used to dissolve native cellulose. ILs is advangeous over other solvents due to the following properties: (1) it is a non-derivatizing solvent that can dissolve cellulose with a degree of polymerization (DP) as high as 650 rapidly without any pretreatment (2) it is an environmentally green solvents for a broad range of chemical and industrial process (3) it can be recycled easily in many cases. Already, 1-butyl-3-methylimidazolium chloride (BmimCl) and 1-allyl-3-methylimidazolium chloride (AmimCl) have been reported to be suitable solvents for cellulose.

However, it should be noted that, the main source of cellulose for application is still cotton and dissolving pulp with high purity by far. This fact makes the raw materials expensive to obtain; on the other hand, the agriculture residue which contains more than 30% cellulose are particularly attractive because of their relatively low cost and plentiful supply. Therefore, the willingness to develop new biomass-based materials, especially the agriculture waste, is very interesting and promising.

In particular, Sugarcane bagasse (SB) as an abundant agricultural lignocellulosic byproduct is a fibrous residue of cane stalks left over after the crushing and extraction of the juice from the sugarcane. Annually, about 54 million dry tonnes of bagasse which contains approximately 50% cellulose is produced throughout the world. But most of this residue is burned, thereby losing energy and causing serious pollution to air, Therefore, this raw material possesses a high economic and environmental potential. So, in the present paper, we have focused on cellulose obtained from agricultural wastes-- SB, which is such a fast growing and cheap sources.

In the present article, we use SB to prepare cellulose, then dissolve and regenerated the sugarcane bagasse cellulose (SBC) in AmimCl solution to obtain regenerated sugarcane bagasse cellulose (RSBC) films.Results showed that the tensile strength of RSBC films prepared in AmimCl can reach as high as 133 Mpa, the value of which is even a bit higher than the RC film prepared from wood pulp at the same condition (128Mpa).

FTIR spectroscopy shows that AmimCl is direct solvent for cellulose; The X-ray results indicate that the change of cellulose I into cellulose II occurred during the dissolution and regeneration process; SEM micrographs of RSBC indicated that after regeneration from AmimCl, the morphology of the dry material is fused into a relatively homogeneous macrostructure from the interior to the surface, displaying a dense texture.

Above all, this study provides a technically feasible and environmentally acceptable method to prepare regenerated cellulose film using relatively cheap sugarcane bagasse as cellulose resources.

See more of this Session: Novel Catalytic and Separation Process Based On Ionic Liquids II
See more of this Group/Topical: Topical G: Innovations of Green Process Engineering for Sustainable Energy and Environment

Numerical Simulation of Single Bubble Rising in Ionic Liquids

Xiaoling Wang,State Key Lab of Multi-phase Complex Systems, Institute of Process Engineering, Chinese Academy of Sciences, Beijing, China

Haifeng Dong, State Key Lab of Multi-phase Complex Systems, Institute of Process Engineering, Chinese Academy of Sciences, Beijing, China

Xiangping Zhang, State Key Lab of Multi-phase Complex Systems, Institute of Process Engineering, Chinese Academy of Sciences, Beijing, China

Suojiang Zhang, State Key Lab of Multi-phase Complex Systems, Institute of Process Engineering, Chinese Academy of Sciences, Beijing, China

Ionic liquids are now wildly used in industrial areas as new technological material. The fluid dynamic of these liquids is one of the most important aspects which are now interest and worthy of study. Predicting the motion of bubbles in dispersed flows is a key problem in fluid mechanics. In this paper, an improved VOF model was proposed and validated for simulation of single bubble rising in three different system, bmimBF4, bmimPF6, omimBF4. In the new model, we considered not only the surface tension proposed by Brackbill et al. , but also the drag force, which were produced between the two phases by a friction pressure drop. The comparison of the simulations with the experiments shows satisfactory. The terminal velocity of the bubble is some smaller than that of the experiment. The effect of bubble diameter on the rise velocity was not evident. Increase of the Eötvös number would increase the bubble velocity, with decreasing Eötvös number, the bubble gets its final diameter, and the equivalent diameter is larger than that of beginning. The velocity and the pressure fields around the bubble are presented (see Fig.1). At the top of the bubble, the liquid is pushed forward and fall around the bubble, there is no radial velocity along the axis far away from the bubble, and high downward liquid velocity is obtained near the bubble. The fluid flow in the region surrounding the bubble leads to a high pressure gradient at the bottom surface of the bubble. As the bubble rises, the upper surface receives a relatively steady flow resistance, which makes the deformation of the bubble.
Acknowledgement: This work was supported by the National Natural Science Funds for Distinguished Young Scholar (No. 20625618).

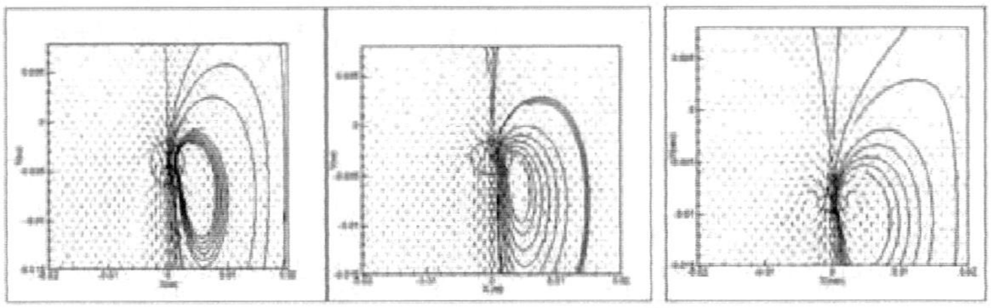

(a) (b) (c)

Fig.1 vectors of velocity magnitude for bmimBF$_4$ at different temperature of (a) 310K, (b) 330K, (c) 355K between 10mm and 33mm above the rector.

See more of this Session: Novel Catalytic and Separation Process Based On Ionic Liquids II
See more of this Group/Topical: Topical G: Innovations of Green Process Engineering for Sustainable Energy and Environment

Experimental Measurements of Amine-Functionalized Ionic Liquids and Carbon Dioxide for Gas Separation

Brett F. Goodrich, Chemical and Biomolecular Engineering, University of Notre Dame, Notre Dame, IN

Lindsay E. Ficke, Chemical and Biomolecular Engineering, University of Notre Dame, Notre Dame, IN

Joan F. Brennecke, Chemical and Biomolecular Engineering, University Of Notre Dame, Notre Dame, IN

Ionic liquids, organic salts with low melting points, have shown great potential to replace the current carbon capture technology which uses an aqueous amine solution. The major problems with aqueous amine solutions are high parasitic energy load, corrosion and degradation in the presence of oxygen. This work focuses on ionic liquids which can chemically complex to the carbon dioxide for gas separation by experimentally measuring the carbon dioxide capacity, heat of absorption, and viscosity.

The capacity of carbon dioxide in ionic liquids was measured using a stoichiometric method at room temperature and pressures between (10 - 1000) mbar. Carbon dioxide uptake is as high as one mole of carbon dioxide per mole of ionic liquid, with tunable capacity by placement of the amine on the anion or cation and choice of substituents.

The heat of absorption of the gas in the ionic liquid was investigated by calorimetry. This gives information regarding the heat of reaction with carbon dioxide, which determines the energy requirement to remove the carbon dioxide from the ionic liquid. We see very low enthalpies for ionic liquids which physically dissolve carbon dioxide and enthalpies which are much higher for the amine-functionalized ionic liquids. However, the enthalpy can be varied significantly by careful choice of substituents.

Another important property of ionic liquid and carbon dioxide mixtures is viscosity. The addition of carbon dioxide to the ionic liquid typically results in a large increase in viscosity, which could hinder industrial applications. Additives such as water, which would be present in the ionic liquid during removal of carbon dioxide from flue gas, were examined to lower the viscosity. The addition of just 0.1 % water by weight showed a 10 % decrease in viscosity, but at a cost of slightly decreased carbon dioxide uptake.

See more of this Session: Novel Catalytic and Separation Process Based On Ionic Liquids II
See more of this Group/Topical: Topical G: Innovations of Green Process Engineering for Sustainable Energy and Environment

Research On H2S Removal and Oxidation From Acidic Gas with [Bmim]2BF4OH

Hongwei Yao, Institute of Process Engineering, Chinese Academy of Sciences, Beijing, China

Xiangping Zhang, Institute of Process Engineering, Chinese Academy of Sciences, Beijing, China

Suojiang Zhang, Institute of Process Engineering, Chinese Academy of Sciences, Beijing, China

Hydrogen sulfide (H_2S), which can cause catalyst poisoning and pipeline corrosion, is produced along with methane in natural gas fields, coal gas fields and crude oil fields. General absorbents of H_2S, such as MDEA, MEA, DGA, DIPA are volatile. In contrast ionic liquids (ILs) provides a vapourless, thermally stable, reusable "green" solvent and catalyst for chemical reaction, so ILs has very possibility for removal of H_2S.

A stable, tack-specific ionic liquids, $[bmIm]_2BF_4OH$, has been used as a catalyst and as a reaction medium in oxidation of hydrogen sulfide (H_2S) to sulfur (S). Very interestingly, although other simple alkaline solvent such as aminoethyl alcohol (MEA), can absorb H_2S, not only can $[bmIm]_2BF_4OH$ improve the absorption of H_2S, but also drive the oxidation of H_2S with oxygen to S exclusively, that is not reported before.

With the ^1H NMR spectra, IR spectra and MS spectra, we can conclude that $[bmIm]_2BF_4OH$ with oxygen forms intermediate of ketone or aldehyde, then is reduced by H_2S, and finally is returned to the original state.

The significant advantages of this strategy are mild reaction conditions (room temperature) and high isolated yields of S (>90%). Because it can combine the absorption and the oxidation, $[bmIm]_2BF_4OH$ shows great potential in green process engineering of oxidative desulphurization of natural gas, coal gas and so on.

Key words: H_2S, Ionic liquid, oxidative desulphurization, Catalysis

Acknowledgments

The authors sincerely appreciate financial supports from National Scientific Funding of China for Distinguished Young Scholar (20625618 and 20806083), National Basic Research Program of China (973 Program, 2009CB219902).

* Corresponding author. E-mail addresses: sjzhang@home.ipe.ac.cn

See more of this Session: Novel Catalytic and Separation Process Based On Ionic Liquids II
See more of this Group/Topical: Topical G: Innovations of Green Process Engineering for Sustainable Energy and Environment

Nano-& Micro-Engineered Adsorption Technology for Rapid CO2 Adsorption/Regeneration

Haiying Wan, Pacific Northwest National Lab, Richalnd, WA

Wei Liu, Energy & Environmental, Pacific Northwest National Lab, Richalnd, WA

CO2 capture from flue gases presents an un-precedent challenge to the separation technology due to low-cost requirement, high gas flow rate, and low head pressure. Capturing CO2 in a short contact time and regenerating the saturated adsorbent rapidly have direct impact on capital and operation costs. A multi-scale material and process engineering approach will be described in this presentation. First, a self-assembly method is used to prepare nano-sized zeolite crystals and enhance intrinsic adsorption/desorption kinetics of the zeolite adsorbent, since the micro-pore diffusion rate has second order dependence on the crystal size. Second, the parent zeolite is modified by ion exchanges to reduce the heat of adsorption and increase the CO2 adsorption capacity, particularly in the presence of moisture. Third, the nano-sized adsorbent powder is incorporated into a micro-channel adsorption bed for practical adsorption/regeneration operation. The micro-structured bed provides high adsorbent loading (>50 wt.%), very low *ƒ*xP for the gas to flow through, little dead or stagnant volume, and fast mass and heat transfer between the channel flow and external surface of the adsorbent particle. Integration of the innovations made in the above three design scales can lead to reduction of the adsorbent bed size by an order of magnitude compared to the conventional packed bed. Scope process design and economics analysis suggests it is possible to reduce the CO2 capture cost by 65% compared to the amine scrubbing process.

See more of this Session: Unconventional Technologies for CO2 Capture, Conversion and Utilization II
See more of this Group/Topical: Topical G: Innovations of Green Process Engineering for Sustainable Energy and Environment

Utilization of Chemical Looping Strategy to Minimize Exergy Loss

Fanxing Li, Chemical and Biomolecular Engineering, The Ohio State University, Columbus, OH

Liang Zeng, William. G. Lowrie Dept. of Chemical and Biomolecular Engineering, The Ohio State University, Columbus, OH

Hyung Rae Kim, William. G. Lowrie Dept. of Chemical and Biomolecular Engineering, The Ohio State University, Columbus, OH

Deepak Sridhar, Chemical and Biomolecular Engineering, The Ohio State University, Columbus, OH

Fei Wang, Chemical and Biomolecular Engineering, The Ohio State University, Columbus, OH

Andrew Tong, Chemical and Biomolecular Engineering, The Ohio State University, Columbus, OH

Nobusuke Kobayashi, Chemical engineering, Nagoya university, Nagoya, Japan

L. -S. Fan, Chemical and Biomolecular Engineering, The Ohio State University, Columbus, OH

A given chemical reaction can be decomposed into multiple sub-reactions through the use of chemical intermediates. A reaction scheme of this nature is denoted as chemical looping reactions. An ideal chemical looping scheme is to design these sub-reactions in such a manner that the exergy (availability) loss in their process applications can be minimized while allowing the separation of the products and/or pollutants generated from the reactions to be accomplished with ease, thereby yielding an overall efficient and economical process system. In this paper, several chemical looping process configurations that convert carbonaceous fuels such as coal and biomass into hydrogen, electricity, and/or liquid fuels are analyzed using exergy analysis. The analysis results indicate that, through the utilization of chemical looping strategy, the irreversibility of the fuel conversion process can be minimized, resulting in significantly improved energy and exergy conversion efficiencies. Effective exergy recuperation techniques that improve the overall energy management of the chemical looping systems are also provided.

See more of this Session: Unconventional Technologies for CO2 Capture, Conversion and Utilization II
See more of this Group/Topical: Topical G: Innovations of Green Process Engineering for Sustainable Energy and Environment

Tailored Synthesis of Precipitated Magnesium Carbonates as Carbon-Neutral Filler Materials During Carbon Mineral Sequestration

Ah-Hyung Alissa Park, Earth and Environmental Engineering and Chemical Engineering, Columbia University, New York, NY

Huangjing Zhao, Earth and Environmental Engineering and Chemical Engineering, Columbia University, New York, NY

Predictions of global energy usage and demand trends suggest that fossil fuels will remain as the main energy sources for the foreseeable future. Unfortunately, the amount of anthropogenic carbon emitted during the energy production is expected to increase, so that the atmospheric CO2 concentration could reach 580 ppm, a threshold value thought to trigger severe climate change, within a mere 50 years, unless action is taken. Thus, reducing carbon dioxide emissions in order to stabilize atmospheric CO2 levels is crucial. The carbon capture and storage (CCS) schemes generally consist of three major steps: CO2 capture, transportation, and sequestration. Currently, the geological storage of carbon dioxide is considered to be the most economical method of carbon sequestration, while mineral carbonation is a relatively new and less explored method of sequestering CO2. The advantage of carbon mineral sequestration is that it is the most permanent and safe method of carbon storage, since the gaseous carbon dioxide is fixed into a solid matrix of Mg-bearing minerals (e.g., serpentine) forming a thermodynamically stable solid product. The current drawback of carbon mineral sequestration is its relatively high cost. Therefore, this study focuses on the tailored synthesis of high-purity precipitated magnesium carbonate (PMC). The physical properties of PMC are controlled to mimic commercially available CaCO3-based filler materials. The effects of pH, reaction time, and reaction temperature on the mean particle size, particle size distribution, and particle morphological structures have been investigated. Finally, the PMC are synthesized in a solution containing carbonic anhydrase.

See more of this Session: Unconventional Technologies for CO2 Capture, Conversion and Utilization II
See more of this Group/Topical: Topical G: Innovations of Green Process Engineering for Sustainable Energy and Environment

CO2 Capture/Sequestration Via Brine and Caustic Material

Yee Soong, National Energy Technology Laboratory, U.S. Department of Energy, Pittsburgh, PA

Robert Dilmore, U.S. Department of Energy, Pittsburgh, PA

Sheila Hedges, U.S. Department of Energy, Pittsburgh, PA

Craig Griffith, U.S. Department of Energy, Pittsburgh, PA

A great deal of concern has been expressed with regard to global climate change and its link to growing atmospheric concentrations of carbon dioxide. Several strategies are under development that will potentially remove CO2 from the atmosphere or decrease CO2 emission. One such strategy involves the capture of CO2 from large point sources (such as fossil fuel-fired power plants) and long-term storage of captured CO2 underground. Carbon dioxide may be sequestered in various geological formations, including depleted oil and gas reservoirs, unmineable coal seams, basalt formations, and deep saline aquifers. However, there are additional options for CO2 sequestration that involve ex-situ carbonation of caustic industrial waste by-products. This work considers capture/sequestration of CO2 from flue gas through promotion of solubility and mineral trapping in Ca, Mg, and Fe-bearing caustic waste streams.

Coal use for domestic electricity generation is by far the largest market for coal in the United States, representing over 80 percent of annual coal consumption. In addition to energy generation, coal-fired steam-electric generation process, also generates sulfur dioxide, nitrogen oxides, and CO2 emissions and solid coal combustion by-products (CCPs). In U.S. alone over 118 million tons of CCPs are generated each year, a total mass comprising fly ash (60%), bottom ash (15%), boiler slag (1%), and flue gas desulfurization ash (24%). At present, only 30 % of the CCP is utilized for some beneficial application, with the remainder managed as waste. Some of the ashes contain high concentration of Ca as well as high alkalinity. These fly ashes may be good candidates for capturing/sequestration of CO2.

Currently 20-30 billion barrels of saline wastewater are produced annually with the production of oil and gas in the USA. About 35% of this acidic wastewater (pH 3 to 5) is treated and discharged to surface water bodies. Many of the gas/oil-field brines have high concentrations of dissolved Ca, Mg, and Fe ions. Carbonate minerals could be formed in the presence of CO2 under certain conditions. This suggests the possibility of using Ca-, Mg-, and Fe-rich saline wastewater for CO2 sequestration via the formation of carbonate minerals.

Over 70 million tons of bauxite residues are generated annually when aluminum is extracted from the principal ore, bauxite. The residue is mostly comprised of iron and titanium oxides, silica, calcium carbonate, and unrecoverable alumina and soda. The pH of the liquid is as high as 13.5, and the solids and solid surfaces also contain high alkalinity. The caustic nature of the residue causes concerns associated with long-term environmental liability. Worldwide, there are about 200 million tons of bauxite residues that are mostly stored in tailings ponds. For decades, the aluminum industry has been investigating options for treating, disposing, and using bauxite residue. However, environmentally and economically sound methods and processes are still elusive. Addition of CO2 to this caustic material will not only result in carbonate mineral formation but will also serve to decrease the pH of the residue, making disposal more environmentally sound. Moreover, bauxite residue (pH 13.5) can serve as an effective source of caustic material to treat acidic oil and gas field wastewater brines (pH 3 to 5) for mineral carbonation. The use of bauxite residue/brine to capture and store CO2 will serve to not only help mitigate the impact of anthropogenic CO2 on global warming but will also help neutralize caustic industrial waste for safe storage and reuse. Thus, the use of caustic industrial by-products/brine mixtures to capture and store CO2 can serve to not only help mitigate the impact of anthropogenic CO2 on global warming but also serve to achieve complete neutralization of the caustic industrial waste for safe storage and potential reuse.

A novel concept that achieves neutralization of caustic industrial waste material through reaction with acidic oil and gas field wastewater brines with subsequent carbonation of CO_2 in a mixed-flow reactor has been explored. Experiments were conducted to determine the CO_2-bearing capacity of reactive mixtures of brine from the Oriskany Sandstone Formation with three caustic industrial by-products: flue gas desulfurization (FGD) spray dryer ash, Class C fly ash sub-bituminous coal combustion by-product, and bauxite residue slurry from the alumina-production process. Reactions were conducted in a closed, well mixed reactor with a 30% CO_2 balanced by N_2 at ambient conditions. Results show linear relationships between caustic by-product addition and CO_2-bearing capacity. FGD spray dryer ash/brine mixtures exhibited higher CO_2 reactivity than those using Class C fly ash (0.76 moles CO_2, at 24 % solids by weight and 0.036 moles CO_2 at 23% solids by weight, respectively). Bauxite residue exhibited moderate capacities in mixtures with higher percent solids (0.34 moles CO_2 in 40% solids bauxite residue slurry).

See more of this Session: Unconventional Technologies for CO2 Capture, Conversion and Utilization II
See more of this Group/Topical: Topical G: Innovations of Green Process Engineering for Sustainable Energy and Environment

Application of Industrial Wastes to Absorption of Carbon Dioxide

Hyun Kyoo Park, Chemical Engineering, Chung Ang University, Seoul, South Korea

Min Hyuk Hwang, Chemical Engineeing, Chung Ang University, Seoul, South Korea

Sun Geon Kim, Chemical Engineering, Chung Ang University, Seoul, South Korea

Industrial wastes that have calcium silicates as main ingredients include steelmaking slag, wasted cement, fly ashes and so on. We separated and/or re-crystallized calcium oxide and silica from these wastes. The concentration of acetic acid as an extract, extraction time, and extraction temperature had significant influence on the composition of the extract or the raffinate. We found that more silica was extracted with lower temperature, higher concentration of acetic acid and shorter extraction time. Under the same conditions of extraction, silica in the slag was more extractable than that in waste cement. In the former, extracted silica was gelated out of the extract while the recrystallized solids on the gel showed very low CO2 absorptivity. On the other hand, porous solids recrystallized from the latter was more active for the absorption. The absorptivity was kept almost constant in their repeated use.

See more of this Session: Unconventional Technologies for CO2 Capture, Conversion and Utilization II
See more of this Group/Topical: Topical G: Innovations of Green Process Engineering for Sustainable Energy and Environment

AUTHOR INDEX

American Institute of Chemical Engineers
3 Park Avenue
New York, New York 10016-5991

CURRAN ASSOCIATES INC
proceedings.com